Sustainable Manufacturing

Sustainable Manufacturing

Edited by
J. Paulo Davim

First published 2010 in Great Britain and the United States by ISTE Ltd and John Wiley & Sons, Inc.

Apart from any fair dealing for the purposes of research or private study, or criticism or review, as permitted under the Copyright, Designs and Patents Act 1988, this publication may only be reproduced, stored or transmitted, in any form or by any means, with the prior permission in writing of the publishers, or in the case of reprographic reproduction in accordance with the terms and licenses issued by the CLA. Enquiries concerning reproduction outside these terms should be sent to the publishers at the undermentioned address:

ISTE Ltd
27-37 St George's Road
London SW19 4EU
UK

www.iste.co.uk

John Wiley & Sons, Inc.
111 River Street
Hoboken, NJ 07030
USA

www.wiley.com

© ISTE Ltd 2010

The rights of J. Paulo Davim to be identified as the author of this work have been asserted by him in accordance with the Copyright, Designs and Patents Act 1988.

Library of Congress Cataloging-in-Publication Data

Sustainable manufacturing / edited by J. Paulo Davim.
  p. cm.
 Includes bibliographical references and index.
 ISBN 978-1-84821-212-1
 1. Production management--Environmental aspects. 2. Manufacturing processes--Environmental aspects. I. Davim, J. Paulo.
 TS155.7.S856 2010
 658.5--dc22

2010003697

British Library Cataloguing-in-Publication Data
A CIP record for this book is available from the British Library
ISBN 978-1-84821-212-1

Printed and bound in Great Britain by CPI Antony Rowe, Chippenham and Eastbourne.

Table of Contents

Preface ix

Chapter 1. Environmental Impact in Micro-device Manufacturing 1
Jong-Leng LIOW

   1.1. Introduction 2
      1.1.1. Sustainability in micro-manufacturing 5
   1.2. Role of LCA 7
      1.2.1. Energy considerations in micro-device
      manufacturing methods 10
   1.3. Energy consideration in micro-manufacturing .. 14
      1.3.1. Mass and energy balance 14
      1.3.2. Minimum work 17
   1.4. Energy consideration in micro-end-milling
   manufacturing 22
      1.4.1. Energy consumption with spindle and
      slide speed variation 23
      1.4.2. Efficiency of the machining process 27
   1.5. Conclusions 28
   1.6. References 29

Chapter 2. Cutting Tool Sustainability 33
Viktor P. ASTAKHOV

   2.1. Introduction 33

2.2. Statistical reliability of cutting tools as
quantification of their sustainability 37
  2.2.1. State of the art 37
  2.2.2. Cutting tool reliability concept 38
  2.2.3. Practical evaluation of tool reliability under
  invariable cutting conditions 41
2.3. Construction of the probability density function of
the tool flank wear distribution with tool test results   50
  2.3.1. Simplified method 50
  2.3.2. Statistical analysis of tool wear curves 52
2.4. Tool quality and the variance of tool life 58
2.5. The Bernstein distribution 59
2.6. Concept of physical resources of the cutting tool. 67
2.7. References 76

Chapter 3. Minimum Quantity Lubrication in Machining 79
Vinayak N. GAITONDE, Ramesh S. KARNIK and
J. Paulo DAVIM

3.1. Introduction 79
  3.1.1. Cutting fluids and problems related
  to cutting fluids 80
  3.1.2. Dry cutting and its limitations 81
  3.1.3. MQL and its performance in machining 81
  3.1.4. Limitations of MQL 83
3.2. The state-of-the-art research for MQL
in machining 84
  3.2.1. Experimental studies on MQL in drilling . . . 84
  3.2.2. Experimental studies on MQL in milling . . . 86
  3.2.3. Experimental studies on MQL in turning . . . 87
  3.2.4. Experimental studies on MQL in other
  machining processes 89
3.3. Case studies on MQL in machining 90
  3.3.1. Case study 1: analysis of the effect of MQL
  on machinability of brass during turning – ANN
  modeling approach 91

  3.3.2. Case study 2: selection of optimal MQL on machinability of brass during turning – Taguchi approach 99
 3.4. Summary 104
 3.5. Acknowledgments 105
 3.6. References 105

Chapter 4. Application of Minimum Quantity Lubrication in Grinding 111
Eduardo Carlos BIANCHI, Paulo Roberto de AGUIAR, Leonardo Roberto da SILVA and Rubens Chinali CANARIM

 4.1. Introduction 111
  4.1.1. Concern about cutting fluids 113
 4.2. Minimum quantity lubrication 114
  4.2.1. Classification and design of MQL systems . . 116
  4.2.2. MQL application in grinding 118
 4.3. Results. 122
  4.3.1. Plunge external cylindrical grinding. 122
  4.3.2. Internal plunge grinding. 146
  4.3.3. Surface grinding. 154
 4.4. Conclusions 169
 4.5. Acknowledgments 170
 4.6. References 170

Chapter 5. Single-Point Incremental Forming 173
Maria Beatriz SILVA, Niels BAY and Paulo A.F. MARTINS

 5.1. Introduction 173
 5.2. Incremental sheet forming processes 174
  5.2.1. Single-point incremental forming. 174
  5.2.2. Incremental forming with counter tool 176
  5.2.3. Two-point incremental forming 177
 5.3. Analytical framework 179
  5.3.1. Membrane analysis 181
  5.3.2. State of stress and strain 182
  5.3.3. Formability limits 185
 5.4. FE background 187
  5.4.1. Modeling conditions 188

5.4.2. Post-processing of results 189
5.5. Experimental 191
   5.5.1. Forming and fracture forming limit diagrams 191
   5.5.2. SPIF experiments 194
5.6. Results and discussion. 195
   5.6.1. Stress and strain fields. 196
   5.6.2. Formability limits 199
5.7. Examples of applications 203
   5.7.1. Sector shower tray 203
5.8. Conclusions 206
5.9. References 206

Chapter 6. Molding of Spent Rubber from Tire Recycling 211
Fabrizio QUADRINI, Alessandro GUGLIELMOTTI,
Carmine LUCIGNANO and Vincenzo TAGLIAFERRI

6.1. Introduction 212
6.2. State of the art of tire recycling 215
6.3. Direct molding of rubber particles. 221
6.4. Experimental results. 225
6.5. Concluding remarks 233
6.6. References 234

List of Authors 241

Index 245

Preface

According to the NACFAM (National Council for Advanced Manufacturing – USA), *sustainable manufacturing* is defined "as the creation of manufactured products that use processes that are non-polluting, conserve energy and natural resources, and are economically sound and safe for employees, communities, and consumers". In other words, sustainable manufacturing is developing technologies to transform materials and products with reduced emission of greenhouse gases, reduced use of non-renewable or toxic materials, and reduced generation of waste. Sustainable manufacturing includes the manufacturing of "sustainable" products (e.g. manufacture of renewable energy) and the sustainable manufacturing of all products.

The purpose of this book is to present a collection of examples illustrating the state of the art and research developments in sustainable manufacturing. Chapter 1 of this book provides the environmental impact on micro-device manufacturing. Chapter 2 contains sustainability aspects of cutting tools. Chapter 3 covers minimum quantity lubrication (MQL) in machining. Chapter 4 contains information on the application of MQL in grinding. Then, Chapter 5 focuses on single-point incremental forming. Finally, Chapter 6 is dedicated to molding of spent rubber from tire recycling.

This book can be used as a textbook for final-year undergraduate engineering students or as a subject on sustainable manufacturing at the postgraduate level. Also, this book can serve as a useful reference for academics, manufacturing and materials researchers, manufacturing, mechanical and environmental engineers, professionals in manufacturing, and related industries. The scientific interest in this book is evident for many important centers of research, laboratories, and universities throughout the world. Therefore, it is hoped that this book will encourage and enthuse others research in this recent field of science and technology.

The Editor would like to acknowledge his gratitude to ISTE-Wiley for this opportunity and for their professional support. Finally, I would like to thank all the authors for their availability for this work.

Chapter 1

Environmental Impact in Micro-device Manufacturing

This chapter deals with the environmental impact of manufacturing used in the production of micro-devices. It focuses on methods to assess the environmental impact and is largely directed toward the Australian manufacturing industry but can be applied to the micro-device manufacturing industry in general.

The relevance of life cycle analysis (LCA) and thermodynamic considerations are presented which form the current trends for analyzing the sustainability of manufacturing.

This chapter concludes with a simple study of varying the spindle and slide speeds in the micro-end-milling production of T-junctions as an example that offers energy optimization in the micro-device manufacturing industry.

Chapter written by Jong-Leng LIOW.

1.1. Introduction

Micro- and nano-technology have been advocated as emerging areas where substantial reduction in energy consumption and greenhouse emissions can be achieved and have been endowed with the prospect of being a candidate for sustainable technological development. In 2002, the U.S. manufacturing industry required a total input of 24,100 PJ [EPA 03] equivalent to 3.94×10^9 barrels of oil, which is more than the total U.S. domestic production of oil (3.11×10^9 barrels) [CIA 08]. Electricity accounts for 40% of the total energy consumed by the U.S. manufacturing industries.

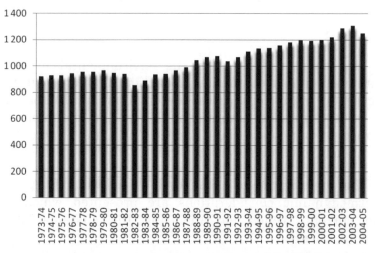

Figure 1.1. *Energy usage (PJ) in the last three decades by the manufacturing sector in Australia [ABA 08]*

In Australia, the manufacturing sector used 1,300.5 PJ in the year 2007–2008. Over the last two decades, its average annual growth rate of 1.4% [ABA 08] has been steady after a decline in 1982–1983, as shown in Figure 1.1. Although manufacturing in Australia consumes 22.5% of the total energy used, it only contributes 10% of the economic output.

The greenhouse equivalent CO_2 (e-CO_2)[1] gas emission, based on the Kyoto accounting method, for the manufacturing and construction industry, is shown in Figure 1.2. The emission represents on average about 9% of the total greenhouse gas emissions for Australia. Although the greenhouse gas emission from the manufacturing and construction industry has increased in the last decade, the overall percentage has been steady due to large increases in emissions from the mining industry. Over the last two decades, the proportion of electrical energy used in the manufacturing industry has increased significantly. Overall, the manufacturing industry is considered an energy-intensive sector with a significant CO_2 footprint.

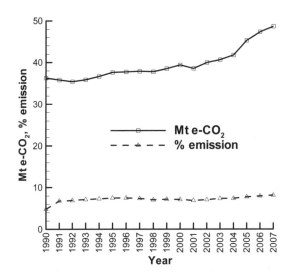

Figure 1.2. *Energy usage in the last three decades by the manufacturing sector in Australia [COM 09a, COM 09b]*

---

1 e-CO_2 includes carbon dioxide (CO_2), methane (CH_4), nitrous oxide (N_2O), perfluorocarbons (PFC), hydrofluorocarbons (HFC), and sulfur hexafluoride (SF_6).

The Australian manufacturing industry consists of more than 75,000 business organizations employing 1 million workers, and changes in the energy competitiveness of the industry can be quite significant for job stability in the manufacturing industry. Manufacturing Skills Australia has recognized the need for the manufacturing industry to respond to climate change, as pointed out by the chief executive of the Australian Industry Group, Heather Ridout [MSA 08]:

Responding to climate change will require a fundamental shift in Australia's approach to management and workforce skills. Reducing greenhouse gas emissions means new processes for industrial and agricultural production, new research and investment in low-emission technologies, new patterns of consumption, and innovative thinking in almost every aspect of business life.

We know little about the carbon footprint and the impact of micro- and nano-technologies on the environment. The U.S. Department of Energy recently released the following claim for nano-manufacturing [USD 09], "Advances in cost-effective nano-manufacturing can deliver diverse energy benefits". A recent analysis of the environmental cost of a single DRAM chip [WIL 02] showed that it had a high-energy intensity and processing material usage, suggesting that the converse is true. This energy-consuming outcome was attributed to the highly organized structure of the DRAM. Similarly, micro-devices manufactured with micro- or nano-technologies have highly organized structure and as one would expect, a high-energy cost per unit. Another important factor is that micro-devices are mostly manufactured with large-scale machines that are inefficient for micro-device manufacturing, a legacy of the techniques brought over from the manufacturing industry. There has been only a few LCAs of micro- and nano-manufacturing processes, and most of the energy efficiency claims have been extrapolated by the belief that anything that is small will use less energy. Thus, micro- and nano-technologies have been advocated as emerging areas where substantial reduction in energy consumption and greenhouse emissions can be achieved and have been endowed with the prospect of being a candidate for sustainable technological development. This premise is currently being challenged by several studies. A more important question is whether it is possible to optimize the manufacture of micro-devices to select energy-efficient routes with minimum environmental impact? This question is in urgent need for an answer with the rise of large-scale micro-device manufacturing.

1.1.1. *Sustainability in micro-manufacturing*

There are several manufacturing methods that are used in the micro-device manufacturing industry. Apart from the standard techniques used in the semiconductor industry, these also include soft lithography on photoresistant polydimethylsulfoxide copies from master templates, focused ion beam, laser micro-machining, water jet milling, micro-electrodischarge machining, LIGA, diamond milling and micro-machining, chemical etching and micro-machining, silicon machining, mechanical micro-machining, and other less common manufacturing methods. An analysis of the energy consumption of a micro-end-milling machine versus a conventional CNC milling machine showed that there are significant energy-reduction advantages with a micro-machine designed specifically for micro-device manufacture, as most of the energy is consumed by moving the slides and associated services rather than in the actual milling process itself [LIO 09].

In micro-device manufacture, the development of sustainable manufacturing requires the disengagement of the link between increased micro-device manufacture and the rise in greenhouse gas emissions. It will require the replacement of old technology and manufacturing processes with new technology that is less energy intensive per unit of manufactured product. This, in turn, will require re-skilling of the workforce to adapt to the change in technology as well as a paradigm shift to a "greener" thinking by both the management and workers.

Sustainable manufacturing in the micro-device industry requires more than just a product-centered approach, whereby only the steps in the making of a product are analyzed in detailed. A more comprehensive approach is to include the complete product development involving not only all the processes that contribute to the development of a product but also the environmental and life cycle cost of the manufacturing equipment. Such an approach is useful when comparing the relative environmental footprint of different micro-device manufacturing methods, all of which have been marketed as environmentally friendly by the manufacturers. It is important to distinguish between the current environmental and energy-efficiency claims regarding micro-devices and the actual environmental and energy costs associated with the manufacture, usage and life of a single micro-device. The disconnection between the product use and the manufacturing and disposal cost is highlighted with the personal computer (PC). When the micro-processor revolution began in the 1970s, it was seen as the revolution that would reduce paper usage, increase work efficiency for the energy used, and lead to smaller and less energy-consuming machines. The present status is an explosion in PC usage, a mounting disposal problem, large energy usage by the semiconductor manufacturing industry, and the generation of large quantities of chemical waste. The PC, similar to the motorcar, is now a permanent fixture in modern life and the focus has now shifted onto making the complete life cycle of the PC much more environmentally friendly and sustainable. Similarly, the continuous rise in the use of micro-devices is predicted and the future society will be more dependent on such devices for its well-being. The challenge is to ensure that not only the manufacturing processes used are sustainable with minimum environmental footprint but also the use and disposal of the micro-devices have minimum environmental impact. This challenge is not easy to accomplish, as energy requirements per unit mass of micro-devices is significantly higher than that for a macro-device and micro-devices may yet pose problems that have not arisen or been identified that awaits their more prevalent use.

1.2. Role of LCA

Life cycle analysis is a management tool for quantitatively assessing the impact of a product on the environment through its complete life cycle. It has been used extensively in the chemical industry [BRE 97], where management of the impact of chemicals has been a major environmental issue since *Silent Spring* was published [CAR 62]. Applications of LCA to micro-device manufacturing need to satisfy the demands of governments, environmental groups, and the public in the areas of waste minimization, recycling, reduced or zero emission, reduced e-CO_2 reduction, and utilization of renewable energy and material resources. A holistic approach must be used if a realistic and genuine effort is made to understand whether micro-device manufacturing is a sustainable form of manufacturing. The current boundary for the LCA study is usually limited to the production of the micro-device itself. This is expected, as stretching out the boundary often requires information that is not easily available. Including waste and emissions as well as the contribution from the life cycle of the equipment used in the manufacture is important when comparing different manufacturing methods because of differences in energy and material impacts. Large-scale equipment, used for production runs, is highly inefficient for prototyping small quantities. Moreover, many of the new products from laboratories lack substantiation when claims are made that a new micro-product or manufacturing method will be sustainable environmentally.

A complete LCA should follow the life cycle of a micro-device, from the extraction of the raw materials used to its final disposal, which includes the material inputs, transportation, energy generation, use, reuse, maintenance or single use, and recycling. The approach to conducting LCA is outlined in the Australian Standards based on the ISO standards. These are as follows:

1. [ISO 98] AS/NZS ISO 14040 "specifies the general framework, principles and requirements for conducting and reporting life cycle assessment studies".

2. [ISO 99] AS/NZS ISO 14041 "specifies the requirements and the procedures necessary for the compilation and preparation of the definition of goal and scope for a LCA, and for performing, interpreting, and reporting a life cycle inventory analysis (LCIA)".

3. [ISO 01a] AS/NZS ISO 14042 "describes and gives guidance on a general framework for the LCIA phase of LCA, and the key features and inherent limitations of LCIA".

4. [ISO 01b] AS/NZS ISO 14043 "provides requirements and recommendations for conducting the life cycle interpretation in LCA or LCI studies". In particular, it provides guidelines on what should be done if information, as well as sensitivity and consistency checks, are incomplete. Examples of life cycle interpretation are provided in its appendices.

Figure 1.3 shows the inter-relationship between the different stages in carrying out an LCA. The scope of an LCA varies with the changing boundaries and although larger and more inclusive boundaries provide a clearer picture of the energy and material usage, the lack of information or the difficulty in isolating all the contributions and losses may make it difficult to achieve. Most of the LCAs conducted are currently at the factory stage where the machining process can be evaluated under known conditions.

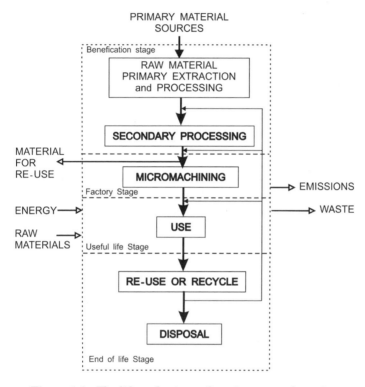

Figure 1.3. *The life cycle stages for micro-manufacturing*

The LCA involves mass and energy balances. Mass balance is often difficult when chemicals are used and re-used in the manufacturing of multiple micro-devices, as the single use of the chemicals is wasteful. Energy balance is more common, as electricity is often a large component of the energy source and can be quantified easily.

1.2.1. *Energy considerations in micro-device manufacturing methods*

The process of micro-device manufacturing consists of several different sub-steps, which are different for every manufacturing process used. In a micro-manufacturing factory, several different manufacturing processes are often used. An LCA of the production of an insert for micro-injection molding [DEG 07] required the use of five major processing steps; plasma vapor deposition, electrical discharge machining (EDM), laser cutting and milling, chemical deposition and dissolution, and cleaning and activation.

The most common approach to the evaluation of the environmental cost of a micro-device manufacturing method is to determine the energy cost for the production of a specific micro-device. As the complexity of the micro-machining required differs from micro-device to micro-device, comparing the energy cost of methods in which micro-devices can be manufactured from start to finish provides a baseline for evaluating the environmental footprint of different micro-machining methods. As the manufacturing can be done in a variety of ways, comparing different manufacturing routes can become a major task, particularly when several machining processes are involved that result in a large number of possible variations.

The energy consumption by machining processes at the production level can often be dominated by static requirements; energy requirements that are inherent in the process irrespective of whether any product is produced. For example, Gutowski *et al.* [GUT 05] showed that the energy use breakdown from a large Toyota machining center identified that a maximum of 14.8% of the energy is used in removing material. This dynamic energy requirement increases linearly from zero to the maximum value as the number of vehicles produced increases. The static energy consumption included cooler, mist collector, etc. (15.2%), oil pressure pump (24.4%), coolant (31.8%), and centrifuge (10.8%). Kordonowy [KOR 01] analyzed several milling machines and found that the energy used in material removal varies from 48% to 69%. The most efficient milling machine was a manual milling machine, as there was less energy requirements from automated auxiliaries (Figure 1.4), whereas the least efficient was an early 1988 Cincinnati Milacron automated milling machine.

The analysis showed that improvements in the energy usage progressed with each newer model of the automated milling facility, resulting in the current suite of newer machines producing as much useful machining work as the older manual milling machines. As seen in Figure 1.4, the work from the specific electrical energy is a linear function of the load; hence, we can write the power (P) per mass rate ($\dot{m}$) as:

$$\frac{P}{\dot{m}} = \frac{P_0}{\dot{m}} + k \qquad [1.1]$$

where P_0 is the idle power and k is a constant.

Similarly, for micro-device manufacturing, the use of conventional CNCs is highly energy inefficient [LIO 09]. The concept of micro-factories whereby micro-components are produced by smaller scale equipment that is designed to handle the small components in possibly clean environments has been studied and implemented in the last two decades [KUS 02, OKA 04]. Micro-factories would occupy less space, provide better environmental control, and consume less energy through a reduction in the energy lost through friction and heat. Thus, the use of micro-milling equipment designed with micro-machining in mind should result in significant energy efficiencies.

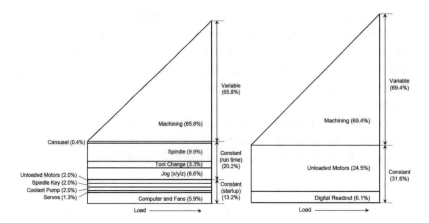

Figure 1.4. *Machining energy breakdown for the 1998 Bridgeport automated milling machine with 5.8-kW spindle motor (left) and the 1985 Bridgeport manual milling machine with a 2.1-kW spindle motor (right) [DAH 04]*

Because of the size of the micro-device, the amount of material removed during the final fabrication of the device is usually small. Most of the materials removed during the process are in the preparatory stages where conventional manufacturing methods are used. In the micro-end-milling of material, the chip sizes that are formed during machining are a few orders of magnitude smaller than those in conventional CNC milling because of the small chip load used, usually of the order of 1 µm. As more exposed surfaces are formed for a given mass of material removed, the energy required to create those surfaces per unit volume of material removed increases. This is similar to the crushing and grinding of ores for which the required energy increases rapidly as the passing screen size decreases. Laser-based techniques are energy intensive, as the material is fully (with CO_2 lasers) or partially (excimer lasers) vaporized, which requires substantial amounts of energy for overcoming the latent heat of vaporization [MOR 07].

In material processing, Gutowski et al. (GUT 09) estimated the electricity requirements for 20 different processes as shown in Figure 1.5. As the process rate decreases, the electricity requirement per kilogram of material increases, with the limits within the following range:

$$\frac{1.19 \times 10^8}{\dot{m}} < w_{elec} < \frac{1.19 \times 10^9}{\dot{m}}, \quad \dot{m} < 10\,\text{kg/h} \qquad [1.2]$$

Given the large amount of electricity used in the micro-device manufacturing processes, it is imperative that alternative processing paths be evaluated to identify low-energy and material usage routes for large-scale manufacturing of a micro-device.

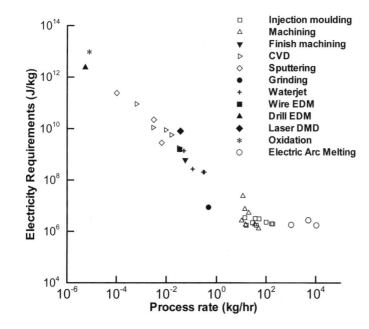

Figure 1.5. *Specific electrical requirement for various manufacturing processes showing the exponentially increasing requirements as process rate decreases (reproduced from [GUT 09])*

1.3. Energy consideration in micro-manufacturing

One of the major procedures in LCA is energy balance. Energy is consumed in manufacturing, most of which is through electricity. The generation of electricity in Australia is largely through coal-fired power stations, and in Victoria, through brown coal, which has a lower energy and higher waste content. Indirectly, it contributes to the depletion of non-sustainable energy source as well as a major contributor of greenhouse gas.

The energy used in the micro-manufacturing process consists of the energy required for processing the materials, energy required for the micro-manufacturing process, and the energy losses during the conversion of energy sources into useful work for the processing. Each aspect can be subdivided into smaller sections to identify where energy is lost. The output is the usefulness of the transformed material. Mass and energy balances provide the means to analyze and thermodynamics will allow us to draw useful conclusions from the energy use and losses for these processes.

1.3.1. *Mass and energy balance*

This analysis closely follows the standard textbook thermodynamics [MOR 08]. The mass balance for a control volume is given as follows:

$$\frac{dm_{CV}}{dt} = \sum_{in} \dot{m}_{in} - \sum_{out} \dot{m}_{out} \qquad [1.3]$$

where m is the mass of the material in a control volume (kg), t is time (s), and $\dot{m}$ is the mass rate in kg/s. The energy balance is:

$$\frac{dE_{CV}}{dt} = \sum_{CV} \dot{Q}_{in} - \sum_{CV} \dot{Q}_{out} - \left(\sum_{CV} \dot{W}_{in} - \sum_{CV} \dot{W}_{out} \right) + \sum_{CV} \dot{m}_{in} h_{in} - \sum_{CV} \dot{m}_{out} h_{out}$$

[1.4]

where E (J) is the energy, $\dot{Q}$ (W) is the heat flow rate, $\dot{W}$ (W) is the rate of work done, and h (J/kg) is the enthalpy of the materials in the control volume. The potential and kinetic energies of the control volumes have been ignored, as the manufacturing system is generally stationary and located on the factory floor.

The entropy balance for a control volume is:

$$\frac{dS_{CV}}{dt} = \sum_{CV} \frac{\dot{Q}_{in}}{T} - \sum_{CV} \frac{\dot{Q}_{out}}{T} + \sum_{CV} \dot{m}_{in} s_{in} - \sum_{CV} \dot{m}_{out} s_{out} + \dot{\sigma}_{CV}$$

[1.5]

where S (J/K) is the entropy change of the control volume, s (J/(kg·K)) is the entropy accompanying the mass flow, T (K) is the temperature, and $\dot{\sigma}$ (W/K) is the rate of entropy production due to irreversible processes within the control volume.

The output is based on the factory conditions, which has a reference temperature of T_0. Rearranging and eliminating the $\sum \dot{Q}_{out}$ term gives:

$$\frac{d\chi}{dt} = \dot{\chi} = \sum_{CV} \dot{W}_{in} - \sum_{CV} \dot{W}_{out} = \sum_{CV} \dot{Q}_{in} - T_0 \sum_{CV} \frac{\dot{Q}_{in}}{T_{in}} + \sum_{CV} \dot{m}_{in} h_{in} - \sum_{CV} \dot{m}_{out} h_{out}$$
$$- T_0 \left(\sum_{CV} \dot{m}_{in} s_{in} - \sum_{CV} \dot{m}_{out} s_{out} \right) - T_0 \dot{\sigma}_{CV}$$

[1.6]

where χ (J) is the exergy of the process and $\dot{\chi}$ (W) is the rate of exergy generated. The exergy provides a maximum theoretical work obtainable from the overall process as it comes into equilibrium with the environment. In calculating the exergy of a manufacturing process in which the material is removed, a large proportion of the exergy flowing into the control volume exits with the product (output). When manufacturing removes a small amount of the material but at a high energy cost per unit volume, the efficiency of the process is not optimized when the amount of starting material is varied, since for the same amount of energy expended, the exergy through the control volume is different. There exists a stream of exergy that flows through the process unaltered and a stream of exergy that is acted upon and active in the control volume considered. The exergy that is acted upon is called the "utilizable exergy" by Sorin *et al.* [SOR 98], and it provides a comparable basis for differentiating the efficiency of different micro-device manufacturing processes.

The efficiency of a process is often defined as:

$$\eta = \frac{\text{Work out}}{\text{Heat input}} = \frac{W_{out}}{Q_{in}} = \frac{W_{out}}{Q_H \left(1 - T_0 / T_H\right)} \qquad [1.7]$$

where the factory environment with temperature T_0 is taken as the minimum temperature at which heat is exhausted. However, for manufacturing processes in which the processing often occurs at the ambient temperature and the products are also at the ambient temperature, there is no work output. Instead, the use of the minimum amount of work required to perform the machining operation to the exergy destroyed provides an efficiency that is much more useful for comparison. Hence, an efficiency of removal as defined by Gutowski [GUT 08] is:

$$\eta_R = \frac{\dot{W}_{min}}{\dot{\chi}_{destroyed}} \qquad [1.8]$$

According to Branham *et al.* [BRA 08], the minimum work input required to effect the transformation of material to product is just the difference between the exergy outputs minus the exergy inputs. The efficiency of the machining process can then be estimated by taking the ratio of the exergy of the useful output and dividing it by the exergy destroyed. Szargut *et al.* [SZA 88] used a slightly different ratio, η_P, calling it the "degree of perfection" defined as:

$$\eta_P = \frac{\chi_{out}}{\chi_{in}} \qquad [1.9]$$

The degree of perfection has been calculated for several different manufacturing processes by Gutowski *et al.* [GUT 09]. Values obtained include 0.7 to 0.9 for injection molding and induction melting of iron, 10^{-4} to 10^{-7} for semiconductor processes, and 10^{-3} to 10^{-7} for processing of nano-materials. The amount of useful work relative to the work input is very small for micro-device manufacture, which does not support the fact that micro-device manufacture is as sustainable as claimed.

1.3.2. *Minimum work*

The calculation of the minimum net work input required to effect the transformation should be independent of the transformation itself. However, every machining process results in a different transformation of the material even if the product is similar, or has similar use and application. Even when only material is removed, the removal process results in different forms of the waste being produced. The least amount of work occurs if the material can be removed unchanged with only cleavage at the planes where the waste separates from the workpiece. This is related to the generation of the minimum amount of new surface area during the process of producing a micro-device. In laser micro-machining, the material is vaporized or ablated from the surface and the extra energy required for melting and vaporizing the material is wasted because the removed material loses all of the energy on refreezing. In electrodischarge machining, energy is also lost through the vaporization of material. In micro-end-mill machining of the material, the removed material is chipped off, resulting in small chips and a large increase in the energy required for creating the extra surface area. Similarly, with water jet machining, the energy abrades the material, similar to grinding, and this incurs a high-energy cost – the efficiency of grinding is at least one or two magnitudes less than CNC machining.

Estimation of the minimum work required for the removal of material in CNC-type machining depends on the models used, and theoretical estimates can be substantially less than the measured values.

1.3.2.1. *Fracture of material*

The first approach is to calculate the minimum work for the creation of a new surface area by fracture. The true surface energy (γ_s) is the work done in creating a new surface area by the breaking of bonds [GUY 72]. For a unit volume of material, equating the true surface energy with the strain energy and assuming that the theoretical tensile strength, $\sigma_T = E/10$, where E is the Young's modulus, gives:

$$\gamma_s = \frac{a\sigma_T^2}{E} \qquad [1.10]$$

where a is the separation between two atom planes. The values of the true surface energy are in the range of 1 to 5 J/m² for brittle material and less than 1 J/m² for plastics and salt. A list of true surface energies for several materials used in micro-machining is listed in Table 1.1. For elastic material, it is necessary to include the effects of plastic deformation and the surface energy is a sum of the true surface energy and the work of plastic deformation (γ_p). The value of γ_p can be increased by suitable changes in the microstructure of the material. In particular, plastics show very large discrepancies between γ_s and γ_p. Guy [GUY 72] provided an example for poly(methyl methacrylate), where $\gamma_p/\gamma_s \approx 1,000$, which means that the minimum work required is increased by three orders of magnitude.

Material	γ_s (J/m²)	Material	γ_s (J/m²)
Metals		*Salts*	
Aluminum	1.1	Calcium fluoride	0.45
Carbon	1.7	Lithium fluoride	0.34
Chromium	2.1	Sodium chloride	0.30
Cobalt	2.7	*Semiconductors*	
Copper	1.9	Gallium phosphide	1.7
Gold	1.6	Gallium arsenide	0.9
Iron	2.9	*Others*	
Nickel	2.5	Alumina	1.4
Platinum	2.7	Glass	0.9
Silicon	1.2	Mica	4.5
Silver	1.3	Perspex	0.5
Titanium	2.6		
Vanadium	2.9		

Table 1.1. *True surface energy for materials used in micro-machining manufacture of micro-devices [GUY 72, COT 75]*

High-strength materials, such as alloys of iron, titanium, and aluminum, are often characterized by low toughness, and the fracture toughness (K_c) is used to associate the material with the fracture stress. The fracture toughness is related to the work of plastic deformation as follows:

$$\gamma_p = \frac{K_c^2}{2E} \qquad [1.11]$$

The work for plastic deformation is likely to yield a more realistic value for the minimum work for the machining of metals. The minimum work obtained is based on the area of new surfaces formed.

1.3.2.2. *Work under the stress-strain curve*

Fracture of ductile material that has undergone appreciable plastic deformation occurs in the necked region. Work is required for the formation of the new surface and the plastic deformation accompanying the removal process. This involves plastic deformation of the area, which may include not only the material that is removed but also the surface that is in contact with the tool. The minimum amount of work can be estimated as the energy required as a result of the plastic deformation. The energy per unit volume is given as follows [GUY 72]:

$$\frac{\int F\,ds}{V} = \frac{W_{min}}{V} = \int_0^{\varepsilon_f} \sigma\,d\varepsilon \qquad [1.12]$$

The integral in the rightmost term is the area under the true stress-strain curve and is much larger for ductile materials than for brittle materials. For a linearly increasing plastic region, Branham et al. [BRA 08] provided the energy per unit volume as follows:

$$\frac{W_{min}}{V} = \sigma_{yield}\left(\varepsilon_{total} - \varepsilon_{yield}\right) + 0.5\left(\sigma_{max} - \sigma_{yield}\right)\left(\varepsilon_{total} - \varepsilon_{yield}\right) \qquad [1.13]$$

The work obtained from the area under the curve provides a larger value of work than the previous estimate from the true surface energy. This approach provides the minimum work per unit volume based on the work required for plastic deformation and hence fracture the material to be removed. Branham et al. [BRA 08] used this method to obtain a value of 4.7×10^7 J/m^3 for the CNC machining of AISI 1212 steel.

1.3.2.3. *Shear energy for material removal*

Metal cutting involves concentrated shear along a rather distinct shear plane. Mechanistic models for steady-state cutting have been based on the processes occurring at the shear plane, starting with the early models of Merchant [ERN 41, MER 44]. In orthogonal cutting [SHA 05], the total energy consumed per unit time is as follows:

$$U = F_p V \qquad [1.14]$$

where F_P is the force in the horizontal direction and V is the velocity of the tool. The specific energy (total energy per unit volume) of material removed is as follows:

$$u = \frac{U}{Vbt} = \frac{F_P}{bt} \qquad [1.15]$$

where b and t are the width and depth of cut, respectively. This specific energy is a sum of the shear energy on the shear plane, the friction energy on the tool face, the surface energy due to the formation of new surface area, and momentum energy due to the momentum change associated with the metal as it crosses the shear plane. The energy balance is dominated by the shear and friction energies, whereas the surface and momentum energies are small and often neglected. Hence, this estimate of the minimum energy required will be substantially larger than the earlier estimates. The shear (u_S) energy and the friction (u_F) energy per unit volume are given as follows:

$$u_S = \frac{F_S V_S}{Vbt}, \quad u_F = \frac{F_C V_C}{Vbt} \qquad [1.16]$$

where F_S is the shear force along the shear plane, F_C is the force along the tool face, V_S is the shear velocity (velocity of the chip relative to the workpiece and directed along the shear plane), and V_C is the velocity along the tool face. The specific energy of cutting can be estimated from published values for a specific material. Shaw [SHA 05] provided a value of 4.914×10^9 J/m^3 for the milling of stainless steel with a continuous chip of thickness of 0.25 mm, with no built-up edge formation and an effective rake angle of 0°. This is an order of magnitude higher than the minimum work per unit volume obtained from the stress-strain curve earlier. Methods for further estimating the shear and friction forces are given by Shaw [SHA 05].

1.4. Energy consideration in micro-end-milling manufacturing

In a study of energy requirements for a micro-end-milling setup, Liow [LIO 09] showed that the bulk of the energy is consumed by the computer as shown in Table 1.2. The results presented are for a particular setup of the micro-machining process – a series of four T-junctions as shown in Figure 1.6. The study used a particular spindle speed (40 krpm), chip load (1 µm per tooth), and depth of cut (5 µm) to machine the T-junctions. During the micro-machining process, the spindle, slides, and the computer required energy whether or not milling was taking place. The static and dynamic parts of the energy requirements were not isolated. Here, a more detailed study of the T-junction milling is presented with an optimization of the energy requirements as a function of the spindle and slide speeds and a comparison with the minimum energy requirement.

Part	Energy consumed (MJ)	% of energy use
Spindle	0.101	19.7
Slides	0.069	13.5
Lubricant flow	0.077	15.0
Airflow	0.015	2.9
Computer	0.251	48.9

Table 1.2. *Energy consumption for a particular case of micro-end-milling T-junctions with a micro-machining setup [LIO 09]*

The T-junction channels are 100 µm wide and 50 µm deep cut into stainless steel. The energy requirements of the spindle and slides are shown in Figure 1.7. The static energy requirements for each slide and the spindle are 7 and 44 W, respectively.

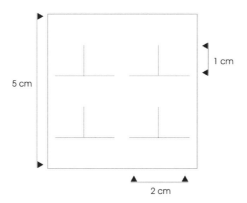

Figure 1.6. *Template of the micro-machined T-junctions [LIO 09]*

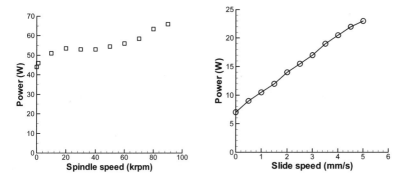

Figure 1.7. *Energy requirements for the spindle (left) and slides (right)*

1.4.1. *Energy consumption with spindle and slide speed variation*

In this study, the spindle and slide speeds were varied. At any given time, only one of the slides is moving while the other two are stationary. The z-axis slide performs minimum movement during the machining of the channels, with the bulk of the work being done by the x and y slides. The cut per tooth changes with the varying spindle and slide speeds. Valid values for cut per tooth range between 0.5 and 5 μm. The lower limit is restricted by the knowledge that at small cut per tooth, the end-mill tends to plough rather than cut. This, in turn, changes the dynamics of the cutting and hence the energy requirements for the cutting process. The surface quality, determined from the average roughness, changes with the different cutting conditions, but it is not considered in this study. A full optimization will require the surface quality to be one of the process constraints. Details of the setup and machining procedures can be found in Liow [LIO 09].

The energy required for micro-end-milling the T-junctions of Figure 1.6 is shown in Figure 1.8. The energy required ranges from 0.046 MJ at 50 krpm and 7.5 mm/s to 0.612 MJ at 50 krpm at 0.5 mm/s. The contours are bounded at the upper left corner by the minimum chip size and at the lower right corner by the maximum chip size. The energy required is independent of the spindle speed and a strong function of the slide speed. This is because the spindle power requirement changes by only a third from rest to 80 krpm. A large proportion of the power required is consumed by the static requirement of the AC-DC and frequency converters used to power the spindle. In contrast, the power requirement of the slides increases by a factor of 3.5, going from rest to the maximum allowable speed of 7.5 mm/s. This is reflected by a steep fall in the power requirement as the slide speed increases. The results show that increasing the cutting rate leads to a decrease in the energy required for the manufacture of the T-junctions, which is consistent with the manufacturing industry move to high-speed milling over the last few decades.

The average energy requirement for the cases studied in Figure 1.8 is given in Table 1.3. The PC, a Pentium 3 microprocessor with a cathode ray tube monitor, used for controlling the milling is the main consumer of energy. The slides have the highest variation since power consumption varies significantly with slide speed.

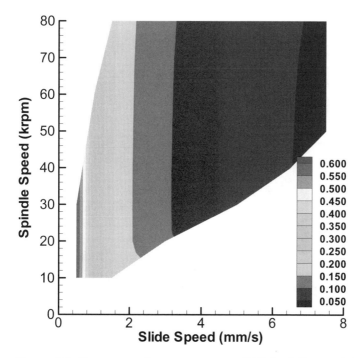

Figure 1.8. *Energy requirement contours (MJ) for different slide and spindle speeds for the micro-end-milling of T-junctions*

Part	Average % total energy	Standard deviation
Spindle	20.91	1.06
Slides	12.55	2.32
Lubricant flow	14.91	0.50
Airflow	2.80	0.09
Computer	48.84	1.64

Table 1.3. *Average percentage of the total energy requirements of the components for micro-end-milling of T-junctions for the cases shown in Figure 1.8*

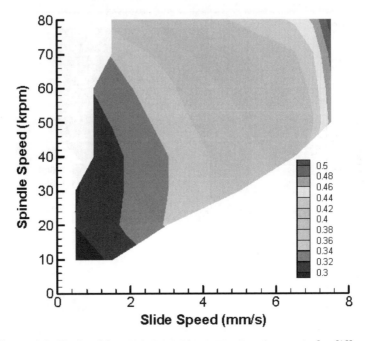

Figure 1.9. *Ratio of dynamic to static energy requirements for different slide and spindle speeds for the micro-end-milling of T-junctions*

The ratio of dynamic to static power requirement is shown in Figure 1.9. A high ratio indicates that the power consumption is due to the cutting process. The static power requirements are quite large and represent 85.2% of the power consumption for the Toyota machining center at full production. The high proportion of static power requirements provides an opportunity to reduce energy consumption independent of the milling process itself. In this case, the increase in dynamic power requirements is a strong function of the slide speed and a weak function of the spindle speed. The ratio increases quite steeply at the top right corner where both spindle and slide speeds are maximized resulting in a significant reduction in the time taken to machine the T-junctions. At the bottom left corner where the milling operation is more work intensive, similar to the grinding process, the time required for milling and the energy used are similarly increased. If the computer is replaced by a more energy-efficient model, it is expected that the ratio of contours will be more pronounced with a larger spread of values.

1.4.2. *Efficiency of the machining process*

The three methods given in sections 1.3.2.1 to 1.3.2.3 are used to calculate the minimum work. The amount of material removed is 5.96×10^{-10} m^3. The minimum surface area removed is 23.76 mm^2 (2.38×10^{-5} m^2).

The true surface energy for stainless steel is approximately 2 J/m^2. From this value, the minimum work required for the formation of the new surface area is 4.75×10^{-5} J. If the surface energy for plastic deformation is used, the surface energy is 4174 J/m^2, given $K_c = 50 \times 10^6$ N/m and $E = 200 \times 10^9$ N/m. This gives the minimum work required as 0.15 J.

The stress-strain curve and equation [1.13], with σ_{yield} = 350 MPa, σ_{max} = 520 MPa, ε_{total} = 0.1, and ε_{yield} = 0.002, give W_{min}/V = 42.63 MJ/m^3. The minimum work is therefore 0.025 J.

Shaw [SHA 05] provided a specific energy value of 4.914×10^9 J/m^3 for the milling of stainless steel with a continuous chip of thickness of 0.25 mm. A correction factor can be applied for the different chip thickness as follows:

$$u \propto \frac{1}{t^{0.2}} \qquad [1.17]$$

Because the chip thickness varies with each different case, the minimum and maximum values of the minimum work calculated by this method ranged from 0.85 to 1.34 J. The efficiencies for the different methods are summarized in Table 1.4.

Method	Minimum work (J)	Efficiency
Surface energy γ_s	4.75×10^{-5}	7.8×10^{-11} to 1.0×10^{-9}
Surface energy γ_p	0.15	2.4×10^{-7} to 3.3×10^{-6}
Stress-strain curve	0.025	4.1×10^{-8} to 5.4×10^{-7}
Specific energy	0.85–1.34	1.4×10^{-6} to 2.9×10^{-5}

Table 1.4. *Efficiency of the micro-end-milling of T-junctions with different approaches to calculating the minimum work*

The efficiencies calculated cover a wide range of more than six orders of magnitudes. In general, these efficiencies are small and there is potential for improving them to make micro-machining manufacture of micro-devices more sustainable.

1.5. Conclusions

The belief that micro-devices are more sustainable because they utilize less energy may not be true particularly during the manufacturing stage. An LCA of a micro-device should include the energy cost of its manufacture, as this stage usually has a high-energy requirement. As the rate of material processed decreases, its energy-processing requirement usually increases. In this study of the micro-end-milling machining of T-junctions in stainless steel, the removal rate was varied between 1.1×10^{-4} and 7.1×10^{-6} kg/h, with an energy requirement between 9.8×10^9 and 1.3×10^{11} J/kg. This high-energy requirement is associated with the low amount of material removed. As the micro-device does not have an inherent capability to do work, the efficiency of the process is given in terms of the minimum work required relative to the work requirement for a given micro-device manufacturing process. Several methods to determine the minimum work have been presented. However, an LCA will require a detailed study of the complete process, which includes manufacturing, use, and recycling.

1.6. References

[ABA 08] ABARE, Energy statistics: 2008, http://abare.gov.au. Accessed October 20, 2009.

[BRA 08] BRANHAM T., GUTOWSKI T., JONES A., SEKULIC D., "A thermodynamic framework for analyzing and improving manufacturing processes", in *Proceedings of IEEE International Symposium on Electronics and the Environment*, San Francisco, CA, 2008.

[BRE 97] BRETZ R., FANKHAUSER P., "Life-cycle assessment of chemical production processes: A tool for ecological optimization", *Chimia*, vol. 51, no. 5, p. 213-217, 1997.

[CAR 62] CARSON R., *Silent Spring*, Houghton Miffin, Boston, MA, 1962.

[CIA 08] Central Intelligence Agency, World Factbook 2008, https://www.cia.gov/library/publications/the-world-factbook/geos/us.html. Accessed October 20, 2009.

[COM 09a] COMMONWEALTH OF AUSTRALIA DEPARTMENT OF CLIMATE CHANGE, *Australian National Greenhouse Accounts: National Inventory by Economic Sector 2007, 2009*, Commonwealth of Australia Department of Climate Change, Canberra.

[COM 09b] COMMONWEALTH OF AUSTRALIA DEPARTMENT OF CLIMATE CHANGE, *Australia's National Greenhouse Accounts: National Greenhouse Gas Inventory Accounting for the Kyoto Target, May 2009, 2009*, Commonwealth of Australia Department of Climate Change, Canberra.

[COT 75] COTTRELL A., *An Introduction to Metallurgy*, Edward Arnold, London, 1975.

[DAH 04] DAHMUS J., GUTOWSKI T., "An environmental analysis of machining" in *Proceedings of IMECE2004 2004 ASME International Mechanical Engineering Congress and RD&D Expo, November 13-19, 2004*, Anaheim, California USA, 2004.

[DEG 07] DE GRAVE A., OLSEN S., HANSEN H., "Life cycle analysis of micro manufacturing process chains – application to the microfactory concept", in *Proceedings of the 3rd International Conference on Multi-material Micro Manufacturing*, Borobets, Bulgaria, 2007.

[EPA 03] U.S. Environmental Protection Agency, Manufacturing footprint: NAICS 311-339. All manufacturing industries 2002, http://www.epa.gov. Published 2003. Accessed October 20, 2009.

[ERN 41] ERNST H., MERCHANT M., "Chip formation, friction and high quality machined surfaces", *Surface Treatment of Metals (Transactions of ASM)*, vol. 29, pp. 299-378, 1941.

[GUT 05] GUTOWSKI T., MURPHY C., ALLEN D., BAUER D., BRAS B., PIWONKA T., SHENG P., SUTHERLAND J., THURSTON D., WOLFF E., "Environmentally benign manufacturing: observations from Japan, Europe and the United States", *Journal of Cleaner Production*, vol. 13, p. 1-17, 2005.

[GUT 09] GUTOWSKI T., BRANHAM M.S., DAHMUS J.B., JONES A.J., THIRIEZ A., SEKULIC D.P., "Thermodynamic analysis of resources used in manufacturing processes", *Environmental Science and Technology*, vol. 43, p. 1584-1590, 2009.

[GUY 72] GUY A., *Introduction to Materials Science*, McGraw-Hill, Tokyo, 1972.

[ISO 98] AS/NZS ISO 14040:1998, *Environmental Management – Life Cycle Assessment – Principles and Framework*. Standards Australia, Homebush, New South Wales, Australia.

[ISO 99] AS/NZS ISO 14041:1999, *Environmental Management – Life Cycle Assessment – Goal and Scope Definition and Inventory Analysis*, Standards Australia, Homebush, New South Wales, Australia.

[ISO 01a] AS/NZS ISO 14042:2001, *Environmental Management – Life cycle Assessment – Life Cycle Impact Assessment*, Standards Australia, Homebush, New South Wales, Australia.

[ISO 01b] AS/NZS ISO 14043:2001, *Environmental Management – Life Cycle Assessment – Life Cycle Interpretation*, Standards Australia, Homebush, New South Wales, Australia.

[KOR 01] KORDONOWY D.N., *A Power Assessment of Machining Tools*. B.Sc. Thesis, Department of Mechanical Engineering, Cambridge, MA, 2001.

[KUS 02] KUSSUL E., BAIDYK T., RUIZ-HUERTA L., CABALLERO-RUIZ A., VELASCO G., KASATKINA L., "Development of micromachine tool prototypes for microfactories", *Journal of Micromechanics and Microengineering*, vol. 12, p. 795-812, 2002.

[LIO 09] LIOW J., "Mechanical micromachining: a sustainable micro-device manufacturing approach?", *Journal of Cleaner Production*, vol. 17, no. 7, p. 662-667, 2009.

[MER 44] MERCHANT M., "Basic mechanics of the metal cutting process", *J. Appl. Mech.*, vol. 11, p. A168-175, 1944.

[MSA 08] MANUFACTURING SKILLS AUSTRALIA, *Sustainable Manufacturing: Manufacturing for Sustainability*, Australian Government, Department of Education, Employment and Workplace Relations, Canberra, Australia, 2009.

[MOR 07] MORROW W., QI H., KIM I., MAZUMDER J., SKERLOS S., "Environmental aspects of laser-based and conventional tool and die manufacturing", *Journal of Cleaner Production*, vol. 15, p. 932-943, 2007.

[MOR 08] MORAN M., SHAPIRO H., *Fundamentals of Engineering Thermodynamics*, John Wiley & Sons, New York, 2008.

[OKA 04] OKAZAKI Y., MISHIMA N., ASHIDA K., "Microfactory – concept, history and developments", *Journal of Manufacturing Science and Engineering*, vol. 126, p. 837-844, 2004.

[SHA 05] SHAW M., *Metal Cutting Principles,* Oxford University Press, Oxford, England, 2005.

[SOR 98] SORIN M., LAMBERT J., PARIS J., "Exergy flows analysis in chemical reactors", *Transactions of IChemE*, vol. 76A, p. 389-395, 1998.

[SZA 88] SZARGUT J., MORRIS D., STEWARD F., *Exergy Analysis of Thermal Chemical and Metallurgical Processes*, Hemisphere Publishing Corporation and Springer-Verlag, New York, 1988.

[USD 09] US DOE. Nanomanufacturing, http://www.eere.energy.gov. Published 2009. Accessed October 20, 2009.

[WIL 02] WILLIAMS E.D., AYRES RU, HELLER M., "The 1.7 kilogram microchip: Energy and material use in the production of semiconductor devices", *Environmental Science and Technology*, vol. 36, 5504-5510, 2002.

Chapter 2

Cutting Tool Sustainability

This chapter discusses practical aspects of cutting tool sustainability. This sustainability stems from tool reliability as its fundamental. Common methodologies for the assessment of cutting tool reliability are presented. It is argued that these common methods use normal, log-normal, and Weibull distribution functions that may not be applicable for tool reliability testing in production environment where the test results are meant to improve cutting tool sustainability of modern unattended machining operations widely used in the automotive industry. To solve the problem, the application of the Bernstein distribution function is discussed using a practical example. The concept and examples of the cutting tool physical resources as the most advanced approach to improve tool sustainability are also presented.

2.1. Introduction

While the term "sustainability" can be used in a variety of ways, the most commonly referenced citation of this term

---

Chapter written by Viktor P. ASTAKHOV.

can be traced to the United Nation's Brundtland Commission. According to the Organization for Economic Co-operation and Development, the term "sustainable development" was introduced in 1980 and popularized in the 1987 report of the World Commission on Environment and Development. In the commission's report, "sustainable development" was defined as: "Development that meets the needs of the present without compromising the ability of future generations to meet their own needs" [UNI 87]. Open to interpretation, "sustainability" has come to encompass as much or as little as is required to fit the needs of its audience. For this reason, the definition requires further clarification to be applicable in manufacturing.

Since its introduction, organizations have tried to apply this concept to industry. One idea in particular, the "triple bottom line", emerged as the business case for sustainability [ELK 98]. This philosophy suggests a more holistic approach that relies on the principles of economic prosperity, environmental stewardship, and corporate responsibility. However, this idea, and the associated terms – "people planet profits", "sustainable management", "ecological sustainability" – all have similar issues. Without metrics to define the achievement of sustainability, success in this arena cannot be measured [ROC 09].

The idea of monetizing sustainability catalyzed the evolution of the triple bottom line into the sustainability ecosystem. These tenets – environmental compliance, communication, and operational efficiency – provide a measurable path forward that is supported by traditional business directives. Increased productivity, reduced plant operating costs, reduction in work effort, and enforced compliance to government regulation have always been driving forces that justify investments in plant optimizations. Taken on its own, each business directive can be related back to more efficient use of necessary resources – energy, raw materials, human resources, information, and equipment – which relates back to a measure of efficiency. Taken together as an optimization strategy, the capability of a solution to meet the immediate needs of the plant has a positive impact on business in the future, not only for the company but also for future generations [ROC 09].

The development of sustainable manufacturing processes should start with simple and measurable building block components. One of the most important components is that the implementation of the reliable cutting tools in manufacturing processes that lead to increased efficiency, reduced cycle times, and reduced human error and potential rework, are all driven by economics. At the same time, these changes reduce energy expenditures, reduce labor – by reducing the use of gasoline consumption and capital expenditures such as office space and the energy required to power and heat them, and minimize scrap material, all facets of environmental stewardship. The data from these efforts have been traditionally used to make decisions on what to produce and when to produce it. But these data can also be used to make further cost-reduction decisions, such as shifting production schedules to accommodate running in off-peak hours and potentially selling surplus energy back to the grid, forwarding the latest trend: corporate responsibility. Therefore, viewed from a different perspective, these types of solutions provide energy conservation and control, building and plant automation, and electrical energy management solutions. In the author's opinion, further exploration of how these conventional systems – and others – meet the goals of sustainability should be related to manufacturing process reliability and dependability.

In performing machining operations on a workpiece in a machine tool, the rate of wear of the cutting tool edge depends largely on the tool and workpiece materials and machining conditions such as feed, speed, and depth of cut. Thus, the rate of wear can vary within wide limits. It is generally considered to be economical in production machining to operate the machine tool to achieve high metal removal rates, which means that the cutting tool wears out rapidly. For example, the prime system objective in drilling is an increase in the drill penetration rate (i.e. in drilling productivity). In all industries, on an average, perishable cutting tools seldom represent more than 8% of the total direct/indirect product manufacturing costs. In CNC machining centers and manufacturing cells where $1.00 is the benchmark, for 2,200 operating hours per year, $1.00 min means an operating cost of $132,000 per year for just one machine (cell). Even factoring in 75% efficiency for loading/unloading, changing tools, and setup, an increase in the penetration rate by 50% amounts to potential annual savings of $24,750 per CNC machining center per year.

In intelligent manufacturing, metal cutting tools usually reach the end of their useful lives via wear mechanisms rather than via an abrupt tool fracture. The accumulated cutting time before a tool should be replaced, even with a fixed cutting task, varies by a large factor, sometimes as large as 3 to 1, because of variability in quality among tools in a given batch and due to even greater variation from one tool batch to another. Great allowable variability of the mechanical and physical properties of the work material commonly found (e.g. in the automotive industry) also makes significant contribution to this variation in tool life. Therefore, if no operator is present to detect the tool failure and replace the tool, it is necessary to schedule tool changes on the basis of the shortest expected tool life of a given tool in the batch. The result is a tool budget as well as cost of labor with the cost of stoppage of the machine far higher than would be necessary if each tool could be used at the end of its own unique useful life.

Unmanned machining centers have been rapidly developed for factory use over the last few decades. At present, various CNC machines, production lines, and manufacturing cells are inherently compatible with unmanned manufacture. However, the principal difficulty in the outcome of automated, unattended operation on an around-the-clock basis is that cutting tools were not developed for this purpose or with this idea in mind. Low tool life, a great scatter in tool life, lack of reliability data, and effective and reliable sensors to monitor the unmanned machining center production systems are major contributors to the problem. In the author's opinion, this problem significantly slows down the outcome of the sustainability concept in modern manufacturing. Worn or fractured tools result in the manufacture of products that are outside specification limits; faulty manufacturing may result in a significant scrap of almost finished parts and cause damage to the machine tool itself, which invariably leads to increased manufacturing cost, loss of manufacturing capacity, and unnecessary use of the energy, materials, and labor.

Some basic and advanced ideas on how the cutting tools should meet the goals of sustainability are explored in the following sections. The idea that a significant difference exists in the laboratory methodologies used for the assessment of cutting tool reliability and those used in practice for sustainable machining processes is discussed.

2.2. Statistical reliability of cutting tools as quantification of their sustainability

2.2.1. *State of the art*

Although the concept of cutting tool reliability has attracted the attention of research in the field over the last 30 years [HIT 79, PAN 78, RAM 78-1, RAM 78-2, RAM 78-3, RAM 78-4, WAN 01], this concept has never been the focus of attention of tool manufacturers and tool end users. Moreover, cutting tool reliability is not normally considered as a tool quality criterion while selecting/ordering tools. This is due to the following reasons:

– Cutting tool quality standards do not contain any references to tool reliability and its characteristics.

– There are a great diversity of opinions and approaches to tool reliability definitions and assessments in the literature.

– Tool reliability tests are expensive and time-consuming. These require highly qualified research teams and high-quality sophisticated equipment.

The major reason, however, is that tool users do not ask for cutting tool reliability data when ordering their tools even when designing new supposedly efficient manufacturing processes and highly automated production plants. Although such data can and should be used for establishing tool replacement strategy, estimating tool cost/tool performance bottom line, and increasing efficiency of machining operations by reducing scrap and tool replacement time, that is, to meet the requirement of lean and sustainable manufacturing, these are rarely accomplished because of the lack of standard procedures and practical methodologies in the use of the tool reliability data to achieve these goals. Tool manufacturers do not include these data in their sales pitch, although general talks about lean and sustainable manufacturing are always there.

2.2.2. *Cutting tool reliability concept*

Reliability predictions are one of the most common forms of reliability analysis. These are meant to predict the failure rate of components and overall system reliability. These predictions are then used to evaluate tool design feasibility, compare design alternatives, identify potential failure areas, trade-off system design factors, and track reliability improvement. Reliability prediction has many roles in the reliability engineering process. The impact of proposed design changes on reliability is determined by comparing the reliability predictions of the existing and proposed designs.

The ability of the tool design to maintain an acceptable reliability level under environmental extremes (changes in work material characteristics, coolant concentration, and flow rate, for example) can be assessed through reliability predictions. Predictions can be used to evaluate the need for environmental control systems installed on the machine and their particular characteristics as well as a need for backup systems and tools on a machine or a production line. A reliability prediction can also assist in evaluating the significance of reported failures. Ultimately, the results obtained by performing a reliability prediction analysis can be useful when conducting further analyses such as an FMECA (failure modes, effects, and criticality analysis), an RBD (reliability block diagram), or a fault tree analysis. The reliability predictions are used to evaluate the probabilities of failure events described in these alternate failure analysis models.

The cutting tool reliability data should be a part of tool quality characteristics and thus should be included in tool drawings as with the tool geometry, surface finish, and other quality parameters if real sustainability of manufacturing process is the objective. The tool users should ask for these data and use them in their calculation of the cost per unit to determine the real efficiency of the cutting tool. It would allow the best tool for the application to be selected.

In general, reliability parameters for the cutting tool are the same as those for any other technical product. These parameters are as follows:

Reliability, $R(t)$, is defined as the probability that the cutting tool experiences no failures during the time interval 0 to t_1, defined as the working time T, given that the component or system was repaired to a similar new condition or was functioning at t_0. As such, the working time T is greater than a certain prescribed time.

$$R(t) = \text{prob}\{T > t\} \qquad [2.1]$$

Failure density is defined as the probability per unit time that the cutting tool experiences its first failure at time t, given that the component or system was operating at time 0.

$$f(t) = -\frac{dR(t)}{dt} \qquad [2.2]$$

Failure rate is defined as the probability per unit time that the cutting tool experiences a failure at time t, given that the component or system was operating at time 0 and has survived to time t.

$$z(t) = -\frac{f(t)}{R(t)} \qquad [2.3]$$

The mean time between failures (MTBF) is a basic measure of reliability for replicable items as cutting tools. MTBF can be described as the time passed before a component, assembly, or system fails under the condition of a constant failure rate. Another way of stating MTBF is the expected value of time between two consecutive failures for repairable systems.

$$\text{MTBF} = \int_0^\infty t \cdot f(t) \, dt \qquad [2.4]$$

The values of these basic characteristics can be assessed using tool testing data for a given cutting tool.

Another indicator of tool reliability commonly used in manufacturing is *tool durability*, defined as the cutting time till a selected tool failure criterion (criteria) is reached. As such, this characteristic is considered to be of a stochastic nature. The assessment of tool durability is carried out by cutting tests assuming that this parameter has a normal distribution. Although such a distribution is normally accepted in many standard durability analyses, many researchers in metal cutting studies found that the durability distribution is not symmetrical [DEL 07]. Rather, its distribution has an asymmetrical left branch. This is particularly true when practical tool testing with significant lot sizes is carried out in a production environment.

2.2.3. *Practical evaluation of tool reliability under invariable cutting conditions*

2.2.3.1. *Model based on log-normal distribution of the average flank wear*

Hitomi *et al.* [HIT 79] and El Wardany and Elbestawi [ELW 97] argued that the distribution of the average flank wear VB_B obeys a normal distribution. Therefore, the probability density function of the tool flank wear distribution $f(VB_B)$ has the following form:

$$f(\overline{V}_B) = \frac{1}{\sqrt{2\pi}\sigma VB_B} \exp\left(-\frac{(\ln VB_B - \mu)^2}{2\sigma^2}\right) \qquad [2.5]$$

where parameters μ and σ are the mean and standard deviations of the log-normal distribution of the average flank wear.

It is know that the average flank wear VB_B is a function of the cutting regime (cutting speed v, feed f, and depth of cut d_w) and the time of cutting t_c, that is,

$$VB_B = \psi(v, f, d_w, t_c) \qquad [2.6]$$

When the parameters of log-normal distribution are given by μ (mean) and σ^2 (variance), the statistical meanings of equation [2.6] are considered as follows [HIT 71]:

$$\mu = E[\ln \text{VB}_\text{B}] = \ln \Psi\left[(v, f, d_\text{w}, t_\text{c})\right] \qquad [2.7]$$

$$\sigma = \text{var}[\ln \text{VB}_\text{B}] = E\left[(\ln \text{VB}_\text{B} - \mu)^2\right] \qquad [2.8]$$

$$\ln \text{VB}_\text{B} \in N(\mu, \sigma^2) \qquad [2.9]$$

In metal cutting testing, the functions in equation [2.6] are represented in the following form:

$$\text{VB}_\text{B} = \Psi(v, f, d_\text{w}, t_\text{c}) = c_0 v^{a_v} f^{a_f} d_\text{w}^{a_d} t_\text{c}^{a_t} \qquad [2.10]$$

where c_0, a_v, a_f, a_d, and a_t are constants determined experimentally by proper methodology (e.g. described by Astakhov [AST 06]). From these constants, we can write the final expression for the probability density function of the tool flank wear distribution as follows:

$$f(\overline{V}_\text{B}) = \frac{1}{\sqrt{2\pi}\sigma \text{VB}_\text{B}} \exp\left(-\frac{\left(\ln \text{VB}_\text{B} - \ln\left(c_0 v^{a_v} f^{a_f} d_\text{w}^{a_d} t_\text{c}^{a_t}\right)\right)^2}{2\sigma^2}\right) \qquad [2.11]$$

Hitomi *et al.* [HIT 71] showed that the probability density function of tool life could be obtained by equation [2.11] as follows:

$$f(t) = \frac{1}{\sqrt{2\pi}(\sigma/a_t)t_\text{c}} \exp\left(-\frac{\left(\ln T_{\text{VB}_\text{b}^*} - \ln t_\text{c}\right)^2}{2(\sigma/a_t)^2}\right) = \phi\left(\frac{\ln T_{\text{VB}_\text{b}^*} - \ln t_\text{c}}{\sigma/a_t}\right) \qquad [2.12]$$

where $T_{\text{VB}_\text{b}^*}$ is the cutting time to reach the flank wear VB_B^*. Tool reliability is defined using equation [2.1] as follows:

$$R(t) = \text{prob}\{T > t\} = \int_t^\infty \frac{1}{\sqrt{2\pi}(\sigma/a_t)t_c} \exp\left(-\frac{\left(\ln T_{VB_b^*} - \ln t_c\right)^2}{2(\sigma/a_t)^2}\right) dt =$$

$$\Phi\left(\frac{\ln T_{VB_b^*} - \ln t_c}{\sigma/a_t}\right)$$

[2.13]

The MTBF is calculated using equation [2.4]:

$$\text{MTBF} = \int_0^\infty t \cdot f(t) dt = T_{VB_b^*} \exp\left(\frac{(\sigma/a_t)^2}{2}\right)$$

[2.14]

and the failure rate is calculated using equation [2.3] as follows:

$$z(t) = -\frac{f(t)}{R(t)} = \frac{1}{(\sigma/a_t)t_c}\left(\phi\left(\frac{\ln T_{VB_b^*} - \ln t_c}{\sigma/a_t}\right) \bigg/ \Phi\left(\frac{\ln T_{VB_b^*} - \ln t_c}{\sigma/a_t}\right)\right)$$

[2.15]

Figure 2.1 shows the comparison of the modeled and experimental results for reliability and failure rate. The machining conditions used to obtain experimental results are as follows: work material – plane carbon steel, dry cutting, cutting speed v = 175 m/min, feed f = 0.2 mm/rev, and depth of cut d_w = 1.5 mm. As is seen, the prediction improves when VB_B increases. In other words, the log-normal model is suitable to predict tool reliability for roughing and semi-roughing machining operations for which the average flank wear VB_B is significant and thus can be accurately measured to construct a reliability model. In finishing operations, where VB_B is rather small or when superhard work materials such as polycrystalline diamond (PCD) and polycrystalline cubic boron nitride (PCBN) are used, this model is not applicable.

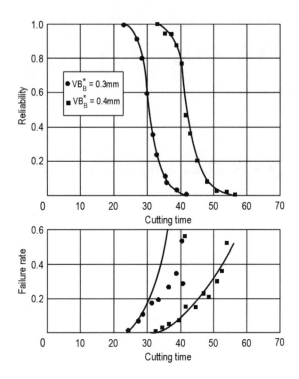

Figure 2.1. *Comparison of the modeled and experimental results for reliability and failure rate*

Figure 2.2 shows the influence of the cutting speed on cutting tool reliability for $VB_B = 0.3$ mm. As is seen, the reliability of the cutting tool decreases with the cutting speed at a much higher rate that the average flank wear. This fact should be accounted for in the design of unattended machining operations. In such a design, the unification of reliability of various cutting tool can be achieved with reduced changing time and lost production time due to tool failures.

The development of a reliability model by the described method requires significant time and human resources; thus, it is costly and specific for a given work material and tool design.

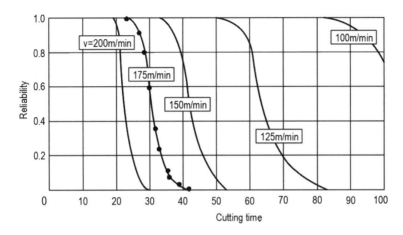

Figure 2.2. *Influence of the cutting speed on tool reliability*

2.2.3.2. *Model based on the reliability-dependent failure rate*

Although there are a number of different approaches to practical assessment of cutting tool reliability by cutting test and measuring the tool wear, the concept presented by Wang *et al.* [WAN 01] appears to be the most reasonable. It considers material properties that are related to the properties of the machining system. This concept is based on Taylor's tool life equation, which correlates cutting speed v with tool life T as follows:

$$vT^n = C \qquad [2.16]$$

where n and C are constants that depend on the properties of the tool and work materials, as well as on the machining regime, and the S/N equation for the fatigue life of a part subjected to alternating stresses. The latter has the following form:

$$SN^m = C \qquad [2.17]$$

where S is the alternative stress, N is the number of fatigue life cycles of a part, m is fatigue exponent dependent on material properties, and C is a constant. Because the success of the implementation of the reliability-dependent failure rate model (AE model) depends on fatigue, sliding wear, and corrosion problems, Wang et al. [WAN 01] extended this model to the assessment of cutting tool reliability.

The AE model includes two decay factors. The first is the so-called decay factor A_0, which associates properties of the cutting tool with properties of the machining system. The second factor is a decay factor A_1, which is related to the properties of the tool material in terms of its wear.

The AE model was developed as follows: If cutting tool reliability decreases monotonically with cutting time (as shown in Figures 2.1 and 2.2), then the reliability function $R(t)$ is a single-valued function of cutting time t_c. Therefore, the probability of tool failure due to tool wear is defined as an analytical function of $[1 - R(t)]$; thus, the failure rate function can be expressed as follows:

$$z(t) = z(1 - R(t)) \qquad [2.18]$$

Equation [2.18] can be represented using Taylor's series expansion as follows:

$$z(t) = A_0 + A_1(1 - R(t)) + \text{higher-order terms} \qquad [2.19]$$

or neglecting higher-order terms [WAN 97] in equation [2.19]:

$$z(t) = A_0 + A_1(1 - R(t)) \qquad [2.20]$$

According to the definition of the failure rate function,

$$z(t) = -\frac{f(t)}{R(t)} = \frac{-dR(t)/dt}{R(t)} = -\left[A_0 + A_1((1 - R(t)))\right]R(t) = A_1\left[R^2(t) - \left(\frac{A_0 + A_1}{A_1}\right)R(t)\right] \qquad [2.21]$$

which provides

$$R(t) = \frac{A_0 + A_1}{A_1 + A_0 \exp\left[(A_0 + A_1)t_c\right]} \quad [2.22]$$

The mean time of failure-free operation (T_0) is as follows:

$$T_0 = \int_0^\infty R(t)\,dt = \frac{1}{A_1}\left(1 + \frac{A_1}{A_0}\right) \quad [2.23]$$

The analysis of equation [2.23] shows that the greater the A_1 is, the severer the decay of reliability of the cutting tool is. Because, as stated previously, A_1 is determined by the cutting conditions, the reliability of a given tool can be improved or adjusted when needed by changing these conditions.

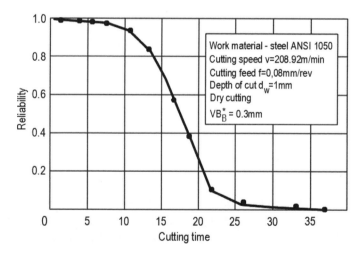

Figure 2.3. *Comparison of the experimental and modeled (using the AE model) reliabilities of the cutting tool*

Figure 2.3 shows the comparison of the experimental and predicted tool reliability. The test conditions are also shown in the figure. The workpiece length is 350 mm and its diameter is 66.5 mm. The standard tool holder MTJNL2525M16 and cutting insert TNMG160404L2G were used in the tests. The average tool flank wear, the standard deviation, and the coefficient of variation related to cutting time were calculated according to the experimental results. The variation in average tool flank wear is shown in Figure 2.4. Taking V = 208.92 m/min as an example, the average tool flank wear, standard deviation, and coefficient of variation are as given in Table 2.1.

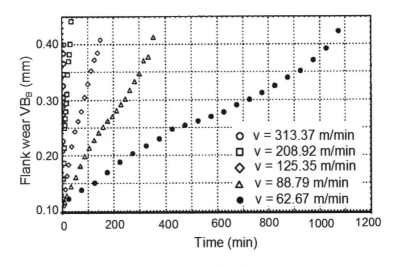

Figure 2.4. *Average tool flank wear for different cutting speeds (cutting feed f = 0.08 mm/rev, depth of cut d_w = 1.0 mm, dry cutting*

Table 2.2 shows the results of the regression analysis used to obtain the average tool life and standard deviation for VB_B = 0.3 mm. The following tool life equation was obtained using these results:

$$vT^{0.322} = 523.029 \qquad [2.24]$$

Tool reliability was calculated using tool life data, and coefficients A_0 and A_1 were obtained by curve fitting. Table 2.3 shows their values for various cutting speeds. As seen, the mean time of failure-free operation (T_0) is more sensitive to the variation in A_1 than to the variations in A_0.

Cutting time (min)	Mean flank wear (mm)	Standard deviation	Coefficient of variation (%)
1.875	0.121	0.0133	10.99
3.750	0.158	0.0173	10.95
5.625	0.186	0.0206	11.08
7.500	0.214	0.0235	10.98
9.975	0.250	0.0255	10.20
11.250	0.255	0.0270	10.94
13.125	0.267	0.0296	11.09
15.000	0.280	0.0312	11.14
16.875	0.295	0.0329	11.15
18.750	0.311	0.0343	11.03
22.500	0.246	0.0372	10.75
26.250	0.370	0.0392	10.59
30.000	0.382	0.0405	10.60
33.750	0.401	0.0441	11.00

Table 2.1. *The mean, standard deviation, and coefficient of variation for tool flank wear test*

Cutting speed (m/min)	Average tool life (min)
313.37	4.90
208.92	17.23
125.35	84.07
88.79	245.11
62.67	722.51

Table 2.2. *Average tool life for different cutting speeds*

Cutting speed (m/min)	$A_0 \cdot \times 10^4$	A_1	rms error (%)
313.37	5.2691	1.6702	0.7822
208.92	2.1235	0.4366	0.9008
125.35	1.8354	0.1004	0.8010
88.79	8.9829	0.0340	1.0288
62.67	3.0525	0.0114	1.0050

Table 2.3. *Parameters of the AE model for various cutting speeds*

An important conclusion can be drawn from the results of the described reliability tests. The spread of the reliability curve along the cutting time axis can be used to evaluate the quality of the cutting tool. The large spread of the reliability curve along the cutting time axis shows relatively low quality of the cutting tool used (3). In other words, if tool life is set at 12 min to ensure 0.9 reliability (which is the most common case for unattended machining operations), then there results a great deal of underused cutting inserts.

The foregoing considerations suggest that a reliability curve should be used in the selection of the proper cutting tool. Unfortunately, this is not the case today because tool users do not ask for such data when considering alternative cutting tool selection. In the author's opinion, sustainability of the cutting tools can be improved if such a curve is available for tool end users.

2.3. Construction of the probability density function of the tool flank wear distribution with tool test results

2.3.1. *Simplified method*

This assessment of tool reliability requires the construction of the probability density function based on tool life testing. The simplest, and most practical, way to obtain this function can be developed if equation [2.5] is represented in the following form:

$$f(T) = \frac{1}{\sqrt{2\pi}\sigma T} \exp\left(-\frac{(\ln T - M(\ln T))^2}{2\sigma^2}\right) \quad [2.25]$$

If tool life test data are available in the form of tool life $T_1, \ldots T_N$, $u_i = \ln T_i$, then their mean u_B and variance σ_u^2 (the parameters of the density function) are calculated as follows:

$$M(\ln T) = u_B = \frac{\sum_{i=1}^{N} u_i}{N} \quad [2.26]$$

$$\sigma_u^2 = \frac{1}{N-1}\sum_{i=1}^{N}(u_i - u_B)^2 \quad [2.27]$$

As an example, consider a log-normal distribution of tool life of a 5-mm diameter high-speed drill with a thick web. The number of tested drills $N = 33$; work material ANSI 4340 steel; cutting speed $v = 16.5$ m/min; and cutting feed $f = 0.095$ mm/rev. A water-soluble coolant of 6% concentration was used. The distribution parameters are as follows:

$$u_B = \frac{\sum_{i=1}^{N} u_i}{N} = 1.53, \quad \sigma_u = \sqrt{\frac{1}{N-1}\sum_{i=1}^{N}(u_i - u_B)^2} = 0.27 \quad [2.28]$$

$$\sigma_u^2 = \frac{1}{N-1}\sum_{i=1}^{N}(u_i - u_B)^2 \quad [2.29]$$

Then, the density function is as follows:

$$f(T) = \frac{0.4343}{T \times 0.27\sqrt{2\pi}} \exp\left(-\frac{(\ln T - 1.53)^2}{0.27^2}\right) \quad [2.30]$$

The distribution function is as follows:

$$F(T) = \Phi\left(\frac{\ln T - 1.53}{0.27}\right) \qquad [2.31]$$

For the degrees of freedom $k = N - 1 = 33 - 1 = 32$, $\chi^2 = 5.0$ and $P(\chi^2) = 0.47$ [SNE 89]. Then, the tool reliability for a 90% confidence level is calculated as follows:

$$P(T) = 1 - F(T) = \Phi\left(\frac{1.53 - \ln T}{0.27}\right) = 0.9 \qquad [2.32]$$

Using the inverse Laplace table [SNE 89], we can find for $\Phi(x) = 0.9$, $x = 1.28$, that is,

$$\left(\frac{1.53 - \log T}{0.27}\right) = 1.28, \text{ then } T_{0.9} = 15.2 \min \qquad [2.33]$$

2.3.2. *Statistical analysis of tool wear curves*

Although the above-mentioned approach is often used to assess tool reliability, it is so oversimplified in the author's opinion that it can lead to significant errors in the assessment results. This is particularly true when a production tool testing method is used to cut testing cost. This is because of the presence of several uncontrolled test parameters. Therefore, a more statistically viable methodology should be used in such an analysis.

2.3.2.1. *Wear curves*

A practical assessment of tool reliability requires a statistical analysis of tool wear curves obtained experimentally. As such, a proper statistical methodology is required to carry out such an analysis properly. This section aims to present an example of the proposed methodology.

In cutting theory and practice, the so-called wear curves [AST 06] are often used to explain phenomenology of tool wear. In full analogy with the classic wear curve, a cutting tool wear curve consists of three distinctive regions, namely, the region of initial wear, the region of steady-state wear, and the region of accelerated wear. The practice of tool wear testing not always confirms this anticipation.

Figure 2.5 shows a fragment of the results of a study dealing with the influence of the threading tap hardness on tool life. As seen, the hardness of the tool material affects not only tool life, which increases with this hardness, but also the appearance of the wear curves. For low hardness, the wear curves are practically straight lines so that there is only one region. For intermediate hardness, the region of accelerated wear becomes obvious so that the wear curves have two distinctive regions. For high hardness, the wear curve resembles the classic wear curve.

Figure 2.6 shows five wear curves (randomly selected out of 30 results) obtained from turning of ANSI 52100 bearing steel. As seen, there are no distinctive regions that can be distinguished on these curves. Moreover, the wear curves shown in Figures 2.5 and 2.6 intersect each other, indicating the stochastic nature of tool wear.

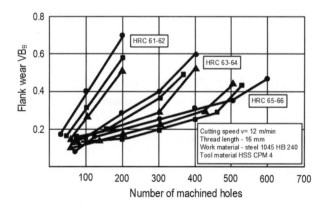

Figure 2.5. *Wear curves for threading taps*

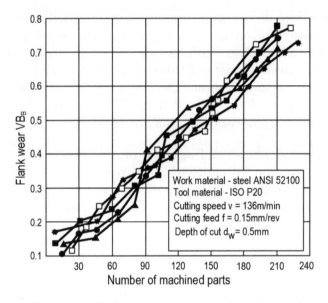

Figure 2.6. *Wear curve for turning of bearing steel*

The foregoing analysis suggests that the appearance of wear curves is not a sufficient indicator for the analysis of cutting tool quality and reliability. Therefore, a more objective statistical methodology should be worked out to help researchers and practical engineers in such an analysis.

2.3.2.2. *Statistical analysis*

A suggested methodology of statistical analysis for the results of tool wear tests is carried out with a practical example. Table 2.4 presents the results of a wear study of an 8-mm diameter M4 high-speed steel twist drill. Table 2.4 shows tool wear increments measured after each hundred drilled holes (14 min of machining time). The total machining time was 84 min, that is, the total number of measurements was $m = 6$. Table 2.5 summarizes the results of calculations.

No. of drill i	Time increment j					
	$t = 14$ min	$2t$	$3t$	$4t$	$5t$	$6t$
1	0.15	0.10	0.10	0.00	0.20	0.15
2	0.20	0.10	0.05	0.05	0.10	0.15
3	0.15	0.10	0.15	0.10	0.10	0.10
...	...	...	...	...	...	...
14	0.15	0.10	0.15	0.20	0.00	0.10
15	0.10	0.00	0.05	0.15	0.10	0.10
16	0.10	0.10	0.05	0.05	0.10	0.00
...	...	...	...	...	...	...
28	0.20	0.15	0.00	0.15	0.00	0.05
29	0.10	0.10	0.00	0.10	0.10	0.05
30	0.10	0.00	0.10	0.00	0.00	0.15

Table 2.4. *Tool wear increments, $\delta w_j^{(i)}$ (mm)*

No. of drill i	$\delta w_B^{(i)}$	σ_i^2	$\log \sigma_i^2$
1	0.117	0.00466	$\bar{3}.6702$
2	0.108	0.00252	$\bar{3}.4014$
3	0.117	0.00068	$\bar{4}.8325$
...	...	...	...
14	0.117	0.00468	$\bar{3}.6702$
15	0.083	0.00268	$\bar{3}.4281$
16	0.067	0.00168	$\bar{3}.2253$
...	...	...	...
28	0.092	0.00742	$\bar{3}.8704$
29	0.075	0.00178	$\bar{3}.2504$
30	0.058	0.00442	$\bar{3}.6454$
$\delta w_B = 0.0903$ mm $\quad \sum \sigma_i^2 = 0.1231 \quad \log \sum \sigma_i^2 = -73.4581$			
$\sum \left(\delta w_B^{(i)} - \delta w_B \right)^2 = 0.010035$			

Table 2.5. *Analysis of wear curves for 8-mm diameter drills*

56 Sustainable Manufacturing

The suggested methodology includes the following steps:

1. For each test drill, the average wear increment gained over time increment $\delta t = 14$ min is calculated as follows:

$$\delta w_B^{(i)} = \frac{1}{m} \sum_{j=1}^{m} \delta w_j^{(i)} \qquad [2.34]$$

For example, for the first drill ($i = 1$):

$$\delta w_B^1 = \frac{1}{6}(0.15 + 0.10 + 0.10 + 0 + 0.20 + 0.15) = 0.117 \text{ mm} \qquad [2.35]$$

2. The overall average wear increment gained over time increment $\delta t = 14$ min is calculated as follows:

$$\delta w_B = \frac{1}{m} \sum_{i=1}^{n} \delta w_B^{(i)} = \frac{1}{30} \times 2.709 = 0.0903 \text{ mm} \qquad [2.36]$$

3. As this step deals with the calculation of variances σ_i^2 of $\delta w_j^{(i)}$, their sum and the average variance of wear increments $\sigma_{\delta w}^2$:

$$\sigma_i^2 = \frac{1}{m-1} \sum_{j=1}^{m} \left(\delta w_j^{(i)} - \delta w_B^{(i)}\right)^2 \quad \sum \sigma_i^2 = 0.1231 \qquad [2.37]$$

$$\sigma_{\delta w}^2 = \frac{(m-1)\sum \sigma_i^2}{mn-1} + \frac{m}{mn-1}\sum_{i=1}^{n}\left(\delta w_B^{(i)} - \delta w_B\right)^2 \qquad [2.38]$$

$$\sum_{i=1}^{n}\left(\delta w_B^{(i)} - \delta w_B\right)^2 = 0.010035 \qquad [2.39]$$

$$\sigma_{\delta w}^2 = \frac{(6-1)0.1231}{6 \cdot 30 - 1} + \frac{6}{6 \cdot 30 - 1} 0.010035 = 0.00377 \qquad [2.40]$$

4. The next step is the verification of the initial assumption that the variances of the observation of the individual groups are equal. This situation is referred to as homogeneity of the variance (and the absence of which is

referred to as heteroscedasticty). The simplest yet most statistically sound method for such a verification is the Bartlett test [BAR 37].

The Bartlett test of the null hypothesis of equality of group variances is based on the comparison of the logarithm of a pooled estimate of variances (across all the groups) with the sum of logarithms of variances of individual groups. The test statistics [SNE 89] applicable for the considered case is based on the calculation of χ^2 represented in the following form:

$$\chi^2 = \frac{2.3026}{1+\frac{n+1}{3n(m-1)}} n(m-1) \left[\log \frac{\sum_{i=1}^{n} \sigma_i^2}{n} - \frac{1}{n} \sum_{i=1}^{n} \log \sigma_i^2 \right] \quad [2.41]$$

For the degrees of freedom $k = n - 1 = 30 - 1 = 29$

$$\chi^2 = \frac{2.3026}{1+\frac{30+1}{3 \cdot 30(6-1)}} 30(6-1) \cdot \left[\log 0.0041 - \frac{-73.4581}{30} \right] = 19.8 \quad [2.42]$$

This value is compared with a value from the table of critical values of χ^2 at a 5% confidence level [SNE 89]. For $k = 29$ and $p = 0.05$, this value is $\chi^2_{cr} = 42$. Because $\chi^2 < \chi^2_{cr}$, the variances of the observation of the individual groups are equal.

5. The next step is to find whether the difference between average increments of the tool wear is significant.

The sum of squares of deviations between testing series:

$$Q_1 = m \sum_{i=1}^{n} \left(\delta w_B^{(i)} - \delta w_B \right)^2 = 6 \cdot 0.01 = 0.06 \quad [2.43]$$

for degrees of freedom $k_1 = n - 1 = 30 - 1 = 29$.

The sum of square deviations within the series:

$$Q_2 = (m-1)\sum_{i=1}^{n}\sigma_i^2 = (6-1)\cdot 0.1231 = 0.616 \qquad [2.44]$$

for degrees of freedom $k_2 = n(m-1) = 30(6-1) = 150$.

The F criterion is then calculated as follows:

$$F = \frac{Q_1/k_1}{Q_2/k_2} = \frac{0.06/29}{0.616/150} = 0.5 \qquad [2.45]$$

This value is compared with a value from the table of critical values of the F criterion at a 5% confidence level, $F_{cr} = 2.2$ [SNE 89]. Because $F < F_{cr}$, the initial assumption (null hypothesis) is valid.

Because the experimental data passed statistical evaluation, they can then be used to determine parameters of the density function. The fitness of the obtained data for normal, log-normal, and gamma distributions is consequently attempted (the test of goodness of fit) and the distribution function that fits best to these data is then used in this reliability analysis.

2.4. Tool quality and the variance of tool life

This section aims to show the difference between tool quality and the variance of tool life, var(T). The simple yet accurate and practical assessment of tool quality gives the value of constant C_v from the experimentally obtained correlation $v = f(T, f, d_w)$ [AST 06]. The variance is the measure of the amount of variation in the tool life with respect to its mean value. Table 2.6 shows data for a twist drill made of the same high-speed steel manufactured by six different tool manufacturers.

Supplier	Relative constant C_{vl}/C_{vi}	Variance var(T)
I	1.00	0.22
II	0.69	0.42
III	0.48	0.47
IV	0.44	0.44
V	0.42	0.15
VI	0.40	0.56

Table 2.6. *Quality and variance of tool life for drills obtained from different tool manufacturers*

As a rule, the higher the quality of tools, the lower the variance of their tool life. However, in the manufacturing practice, there are several deviations from this general rule that allows us to distinguish the quality of tool design including the suitability of the tool material and the quality of tool manufacturing. As shown in Table 2.6, the variance of the tool life for drills produced by tool manufacturer V clearly indicates good manufacturing quality of the drill. The detailed analysis of these tools showed that the flank angle of the lips assigned by the tool drawing was insufficient for the application that caused relatively low tool life whereas the manufacturing process stability and inspection practices were the best among considered drill manufacturers.

2.5. The Bernstein distribution

The major issue with cutting tool testing and implementation is significant variation in their quality even within the same production batch that results in great scatter in tool life. Commonly, only a few cutting tools are used in laboratory testings carried out by universities or R&D departments. These tools are carefully selected from the same production batch, measured, and calibrated.

Although such testing is suitable when attempts are made to study the influence of a particular tool design, geometry features (e.g. the flank angle), or machining regime (e.g. the cutting speed) on the outcomes of the machining process such as tool life, cutting force and power, machining quality, productivity, efficiency, etc., it is completely unacceptable in reliability study results that are to be used for assigning tool lives in unattended manufacturing operations (production lines and manufacturing cells), such as in the automotive industry where a large number of tools are used in mass production of parts.

To obtain adequate reliability results, a large number of tools should be tested. As such, all real-world imperfections such as the difference in machines, coolant supply and quality, variations in part fixtures, tool holders, controllers, etc., should be included in the tests. This makes it difficult to ensure the statistical viability of the test result and even more difficult to assign proper tool reliability for an unassisted machining process, particularly in the automotive industry if the known "classic" distribution function is used as suggested by the known studies on tool reliability. In the author's opinion, the Bernstein distribution should be used to evaluate experimental data for reliability testing in machining [KAT 74]. This section presents a simple yet accurate and practical methodology for test data evaluation.

Consider a simple linear wear represented by wear curve 1 shown in Figure 2.7. Normally, it is approximated by a linear function $w(t) = a(t) + b$, where b is the initial wear, that is, $b = w(0) = w_0$, and a is the wear rate. It is understood that a is a random variable that depends on the quality of a particular tool because this quality directly affects the tool wear rate. If a and b have normal distribution, then $a(t)$ is distributed normal with the following parameters:

$$M\{w(t)\} = M\{a\}t + M\{b\} \qquad [2.46]$$

$$\text{var}\{w(t)\} = t^2 \text{var}\{a\} + \text{var}\{b\} \qquad [2.47]$$

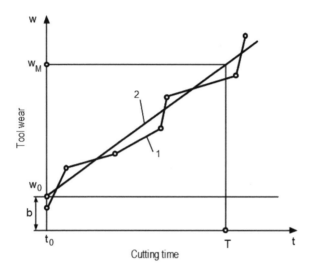

Figure 2.7. *Wear curves: (1) experimental; (2) approximated for statistical analysis*

If w_M is the maximum allowed tool wear that follows from the normality of $w(t)$ and from:

$$P\{t > T\} = P\{(aT + b) \le w_M\} \qquad [2.48]$$

then:

$$P\{t \le T\} = 1 - P\{t > T\} = \Phi\left[\frac{T - \dfrac{w_M - M\{b\}}{M\{a\}}}{\sqrt{\dfrac{\text{var}\{a\}T^2 + \text{var}\{b\}}{M\{a\}}}}\right] \qquad [2.49]$$

This distribution is known as the Bernstein distribution [BOB 02]. It differs from the normal distribution in that *the variance is time dependant*. The Bernstein distribution has three parameters:

$$\alpha = \frac{\text{var}\{a\}}{M^2\{a\}}, \quad \beta = \frac{\text{var}\{b\}}{M^2\{a\}}, \quad \delta = \frac{w_M - M\{b\}}{M\{a\}} \qquad [2.50]$$

Substituting these parameters in equation [2.48] and rearranging terms, we can obtain:

$$F(T) = \Phi\left[\frac{T-\delta}{\sqrt{\alpha T^2 + \beta}}\right] \qquad [2.51]$$

Tool reliability, that is, the probability of the work without failures over operational time T is than calculated as follows:

$$R(t) = P\{t > T\} = 1 - F(T) = 1 - \Phi\left[\frac{T-\delta}{\sqrt{\alpha T^2 + \beta}}\right] \qquad [2.52]$$

The parameters of the Bernstein distribution defined by equation [2.51] can be determined using tool wear curves similar to curve 1 shown in Figure 2.7. To determine these parameters, the following steps are recommended [KAT 74]:

1. A tool wear test is carried out WITH N test tools. The wear curve is drawn for each tool.

2. Each wear curve obtained experimentally is approximated using a least-square method to a straight line (curve 2 in Figure 2.7). Coefficients a_i and b_i are determined for each line.

3. Parameters of the Bernstein distribution are calculated using equation [2.49].

4. Tool reliability is calculated using equation [2.51].

As an example of the Bernstein distribution in tool testing, consider the results of the tests of 10.5-mm diameter twist drills shown in Figure 2.8. The split point grind to ensure self-centring ability of the drill and TiCN coating to ensure improved tool life were applied. Standard manufacturing, inspection, packing, and drill pre-setting procedures were used.

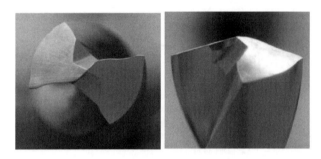

Figure 2.8. *Drills used in the test*

In the author's opinion, the major source of the observed scatter in tool life of the tested drills is the so-called web eccentricity/lip index error. Web eccentricity/lip index error allowed by DIN 1414 or equivalent USCTI tolerance is 3°. Figure 2.9 illustrates how this allowed drill web eccentricity of lip/flute index affects the desired 50/50% chip load (force) balance. While USCTI and DIN standards used the lip height to lip height measurement to determine point eccentricity, 3° and equivalent USCTI decimal allowance have a direct effect on this measurement even if the point is truly concentric.

The problem with this error is that there is no adequate metrological support for the detection of this error. Common inspection equipment locates a drill in a V-block and not on its rotational axis and thus does not allow drill precision rotation about its true longitudinal axis. Common tool pre-setting machines allow focusing only on one lip and then rotating the drill to focus on the other. As such, the web eccentricity cannot be detected. Common tool geometry measurement machines do not include this feature in their basic programs.

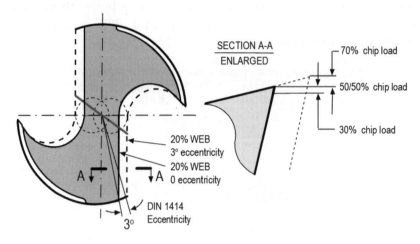

Figure 2.9. *Effect of the allowed drill web eccentricity of lip/flute index error on the chip load (force) balance*

Initial wear w_0 was determined as that reached for $t_0 = 9$ min of drilling, which corresponded to 50 drilled holes. This time period corresponds approximately to the region on initial tool wear on wear curves. The maximum allowed tool wear was $w_M = 0.5$ mm. The drilling time to achieve this wear is considered as tool life T. Tool wear rate is calculated as follows:

$$r_w = \frac{w_M - w_0}{T - t_0} \qquad [2.53]$$

The experimental data and calculations of the tool wear rate are shown in Table 2.7. The mean and variation in the wear rate and initial wear are then calculated as follows:

Drill no.	Initial tool wear w_0 (mm)	Tool life (min)	Wear rate r_w
1	0.10	117.0	0.0037
2	0.07	126.0	0.0037
3	0.07	93.6	0.0051
...	...	...	...
25	0.07	99.0	0.0048
26	0.05	93.6	0.0053
27	0.05	118.8	0.0041
...	...	...	...
43	0.10	77.4	0.0058
44	0.05	126.0	0.0038
45	0.08	77.4	0.0061
	$M\{w_0\} = w_{0B} = 0.08$		$M\{r_w\} = r_{wB} = 0.0054$

Table 2.7. *Results of testing of 10.5-mm diameter drills*

For the wear rate:

$$M\{r_w\} = \frac{\sum_{i=1}^{N} r_{wi}}{N} \quad [2.54]$$

$$\text{var}\{r_w\} = \frac{1}{N-1}\sum_{i=1}^{N}\{r_{wi} - M\{r_{wi}\}\}^2 \quad [2.55]$$

For the initial wear:

$$M\{w_0\} = \frac{\sum_{i=1}^{N} w_{0i}}{N} \quad [2.56]$$

$$\text{var}\{w_0\} = \frac{1}{N-1}\sum_{i=1}^{N}\{w_{0i} - M\{w_{0i}\}\}^2 \qquad [2.57]$$

The results of calculation shown in Tables 2.7 and 2.8 are then used to calculate the parameters of the Bernstein distribution as follows:

$$\alpha = \frac{\sigma_r^2}{r_{wB}^2} = \frac{0.0000035}{0.0054^2} = 0.12 \qquad [2.58]$$

$$\beta = \frac{\sigma_w^2}{r_{wB}^2} = \frac{0.00112}{0.0054^2} = 38.4 \qquad [2.59]$$

$$\delta = \frac{w_M - w_{0B}}{r_{wB}} = \frac{0.5 - 0.008}{0.0054} = 77.7 \qquad [2.60]$$

Drill no.	Variance of the wear rate		Variance of the initial wear	
	$(r_{wi} - r_{wB})$	$(r_{wi} - r_{wB})^2$	$(w_{01} - w_{0B})$	$(w_{01} - w_{0B})^2$
1	0.0017	0.00000289	0.02	0.0004
2	0.017	0.00000289	0.01	0.0001
3	0.003	0.00000009	0.01	0.0001
...	...	...	...	...
25	0.0006	0.00000036	0.01	0.0001
26	0.0001	0.00000001	0.03	0.0009
27	0.0013	0.00000169	0.03	0.0009
...	...	...	...	...
43	0.0004	0.00000016	0.02	0.0004
44	0.0016	0.00000256	0.03	0.0009
45	0.0007	0.00000049	0.00	0.0000

Table 2.8. *Statistical calculations*

The distribution function is calculated using equation [2.51] as follows:

$$F(T) = \Phi\left[\frac{T - 77.7}{\sqrt{0.12T^2 + 38.4}}\right] \qquad [2.61]$$

For the degrees of freedom $k = N - 1 = 45 - 1 = 44$, $\chi^2 = 3.09$ and $P(\chi^2) = 0.22$ [SNE 89]. Then, the tool reliability at 90% percent confidence level is calculated as follows:

$$0.9 = 1 - \Phi\left(\frac{T_{0.9} - 77.7}{\sqrt{0.12T_{0.9}^2 + 38.4}}\right) \qquad [2.62]$$

Using the inverse Laplace table [SNE 89], we can find that $T_{0.9} = 53$ min.

2.6. Concept of physical resources of the cutting tool

Statistical evaluations of cutting tool reliability suffer from system dependency. In other words, a particular assessment is valid, in general, for the particular work material, range of the machining regime, and machining system used for such assessment. A little alteration in these conditions may cause significant variation in the determined tool reliability. This is particularly important in the automotive industry where a large allowable variation in the properties of the work materials keeps down the production costs [AST 06] while high tool reliability is required to run efficiently unattended machining operations in production lines and manufacturing cells. Therefore, another approach to tool reliability should be found.

Analyzing the energy flows to the zone of the fracture of the layer being removed through the cutting wedge (defined as a part of the tool located between the rake and the flank contact areas), Astakhov [AST 06] concluded that out of three components of the cutting system, namely, the cutting tool, the chip, and the workpiece, the only component that has an invariable mass of material and which is continuously loaded during the cutting process is the cutting wedge. As such, the overall amount of energy, which can be transmitted through this wedge before it fails, is entirely determined by the physical and mechanical properties of the tool material.

On the contrary, the material of the chip is not subjected to the same external force because the chip is an evergrowing component, that is, a new section is added to the chip during each cycle of chip formation whereas "old" sections move out of the tool-chip interface and then leave the tool-chip interface and thus do not experience the external load. The same can be said about the workpiece on which volume and thus mass changes during the cutting process as well as the area of load application are imposed by the cutting tool.

When the cutting wedge looses its cutting ability because of wear or plastic lowering of the cutting edge (creep), the work done by the external forces that cause such a failure is regarded as the critical work. As was established by Huq and Celis [HUG 02], a direct correlation exists between wear and the dissipated energy in sliding contacts. Thus, for a given cutting wedge, this work (or energy) is a constant value. The resource of the cutting wedge, therefore, can be represented by this critical work.

According to the principle of physical theory of reliability [KOM 02], each component of a system initially has its resource and this resource is spent during operation time at a given rate depending on the operating conditions. This principle is valid for a wide variety of operating conditions providing that changes from one operating regime to another do not lead to any structural changes in materials' properties (reaching the critical temperatures, limiting loads, chemical transformations, etc.). As such, the resource of a cutting tool, r_{ct}, can be considered as a constant that does not depend on a particular way of its consumption, that is,

$$r_{ct} = \int_0^{\tau_1} f(\tau, R_1) d\tau = \int_0^{\tau_2} f(\tau, R_2) d\tau \qquad [2.63]$$

where τ_1 and τ_2 are the total operating time on the operating regime R_1 and R_2, respectively, till the resource of the cutting tool is exhausted.

The initial resource of the cutting tool can be represented by this critical energy, and the flow of energy through the cutting tool exhausts this resource. The amount of energy that flows through the cutting tool depends on the energy that is required to separate the layer being removed, which, in turn, defines the total energy U_{cs} required by the cutting system to exist [AST 06]. Therefore, there should be a strong correlation between a parameter (or metric) characterizing the resource of the cutting tool (e.g. flank wear VB_B) and the total energy U_{cs}.

To demonstrate the validity of the discussed principle, a series of cutting tests (bar longitudinal turning) was carried out. The work material was AISI 52100 steel: chemical composition: 0.95% C, 1.5% Cr, 0.35% Mn, and 0.25% Si; tensile strength, ultimate, 689 MPa; tensile strength, yield, 558 MPa, annealed at 780°C to hardness 192 HB. Cutting tool material, carbide P10 (cutting inserts SNMG 120408).

The experimental results are shown in Table 2.9. It follows from this table that there is a very strong correlation between the total work required by the cutting system and the flank wear. This correlation does not depend on a particular cutting regime, cutting time, and other parameters of the cutting process. Figure 2.10 shows the correlation curve.

Test no.	Feed (mm/rev)	Depth of cut (mm)	Operating time (s)	Flank wear VB_B (mm)	Energy of the cutting system (KJ)
1	0.07	0.1	8,540	0.45	0.88
2	0.07	0.1	6,680	0.41	0.63
3	0.07	0.1	4,980	0.39	0.52
4	0.07	0.1	1,640	0.29	0.25
5	0.07	0.1	9,120	0.45	0.91
6	0.07	0.1	7,660	0.42	0.68
7	0.07	0.1	6,260	0.41	0.55
8	0.07	0.1	4,900	0.37	0.41
9	0.07	0.1	3,450	0.35	0.47
10	0.07	0.1	5,380	0.38	0.65
11	0.07	0.1	4,240	0.34	0.35
12	0.07	0.1	3,150	0.30	0.33
13	0.07	0.1	2,075	0.26	0.24
14	0.07	0.1	1,036	0.20	0.15
15	0.07	0.1	2,980	0.37	0.37
16	0.07	0.1	1,940	0.32	0.30
17	0.07	0.1	1,190	0.27	0.17
18	0.09	0.1	938	0.15	0.05
19	0.09	0.1	1,874	0.18	0.10
20	0.09	0.1	2,840	0.22	0.22
21	0.09	0.1	3,820	0.25	0.17
22	0.09	0.1	4,810	0.31	0.39
23	0.12	0.1	775	0.20	0.08
24	0.12	0.1	1,520	0.23	0.18
25	0.12	0.1	2,350	0.26	0.29
26	0.12	0.1	3,220	0.27	0.30
27	0.14	0.1	675	0.20	0.18
28	0.14	0.1	1,315	0.21	0.12
29	0.07	0.2	1,295	0.21	0.13
30	0.07	0.2	2,610	0.27	0.30
31	0.07	0.2	5,420	0.44	0.82
32	0.07	0.8	1,316	0.20	0.19

Table 2.9. *Conditions of tests and experimental results*

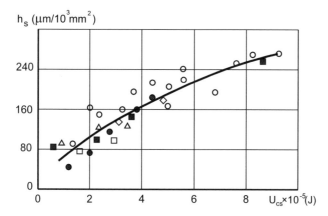

Figure 2.10. *Correlation curve*

The correlation between the energy through the cutting wedge and its wear can be used for the prediction of tool life and optimal cutting speed, allowing the avoidance of expensive and time-consuming tool life tests. Moreover, the multiple experimental results, obtained from machining of different work materials with different cutting tools, proved that this correlation holds regardless of the particular manner the resource of the tool was spent.

The essence of the method can be described as follows: The energy required by the cutting system during the time period corresponding to tool life T_{ct} can be represented as follows:

$$U_{cs} = W_{cs} T_{ct} \qquad [2.64]$$

where W_{cs} is power (in W) required by the cutting system.

Note that U_{cs}, when selected for a given tool material by using the correlation curve similar to that given in Figure 2.10 for the accepted tool life criterion, is the sole characteristic of the tool material, that is, its resource can be used for the calculation of tool life in cutting of different work materials.

The power of the cutting system, W_{cs}, is determined as a product of the power component of the cutting force, F_z, and the cutting speed, v, that is,

$$W_{cs} = F_z v \qquad [2.65]$$

In turn, the cutting force F_z can be determined experimentally depending upon the cutting parameters as follows:

$$F_z = C_{P_z} d_w^{n_z} f^{m_z} v^{k_z} \qquad [2.66]$$

where C_{P_z} is a constant for the work material, and $n_z, m_z,$ and k_z are exponents.

Substituting equations [2.65] and [2.66] into equation [2.64] and expressing tool life, we can obtain an equation that determines tool life for a given cutting regime

$$T_{ct} = \frac{U_{cs}}{C_{P_z} d_w^{n_z} f^{m_z} v^{k_z+1}} \qquad [2.67]$$

If it is necessary to know the cutting speed of the corresponding desired tool life, then equation [2.67] can be expressed as follows:

$$v = \left(\frac{U_{cs}}{C_{P_z} d_w^{n_z} f^{m_z} T_{ct}} \right)^{\frac{1}{k_z+1}} \qquad [2.68]$$

It is obvious that U_{cs} selected on the basis of the tool flank wear depends not only on the properties of the tool material but also on the tool geometry. Therefore, the correlation curve $U_{cs} = f(VB_B)$ should be corrected, accounting for the particular tool geometry. As a result, there are countless numbers of possible combinations "cutting tool material-tool geometry" to account the influence of the tool geometry. To avoid the the influence of tool geometry, the volumetric or mass tool wear, m_v [AST 06], can be used instead of VB_B.

The results of foregoing analysis suggest that the most prospective way to achieve repeatability of cutting tool with inserts is the certification of cutting inserts of standard shapes. The number of standard shapes of cutting inserts (including their geometry) is relatively small, so each insert producer should be able to provide a correlation curve $U = f(m_v)$ for each shape and tool material. Table 2.10 presents some correlations for different tool materials and for different shapes of cutting inserts, obtained experimentally using basic groups of the work material (low carbon medium- and high-carbon steels, low and high alloys including chromium- and nickel-based titanium alloys). The obtained correlation curves do not depend on the particular work material, machine, or any other cutting conditions, so they are properties of the considered tool materials.

Tool material	ISO code of the shape	Correlation curve	Critical temperature (°C)
TS332 (99% Al_2O_3 + 1% MgO, 2,300 HV)	SNMN 120404M	$U = \exp(9.6 VB^2)$	1,200
VOK60 (60% Al_2O_3, 40%TiC, 94 HRA)	SNMN 120404M	$U = \exp(10.91 VB^2)$	1,200
Silinit-P ($Si_3N_4 + Al_2O_3$, 96 HRA)	SNMN 120404M	$U = 573 VB^2$	1,200
TN20 (75% TiC, 15% Ni, 10% Co, 90 HRA)	SNMN 120404M	$U = 434.46 \times 10^{-3} VB^2$	780
Kiborit (96% CBN, KNH 32–36 GPa)	RNMM 1200404M	$U = 50 VB^{1/2}$	1,400

Table 2.10. *Correlation curves for some tool materials*

The correlations in Table 2.10 were obtained for $VB_B \leq 0.4$ mm. The data presented in Table 2.10 are valid under the condition that the tool material does not lose its cutting properties because of excessive temperature. For example, the data for Silinit-P is valid if the cutting temperature does not exceed that given for cutting of steel with feed $f = 0.07$ mm/rev, depth of cut $d_w = 0.1$ mm, and cutting speed $v = 3.3$ m/s. If the cutting speed is increased to $v = 4$ m/s, this material looses its cutting ability, so the correlation curve presented in Table 2.10 is no longer valid.

It has to be pointed out that the limiting work is an complex integral index of the cutting tool resource and intensity of its "spending". This work is not the same for two geometrically alike cutting inserts made of different tool materials because their application for the same work material at the same machining regime results in different works of chip formation defined by the chip compression ratio [AST 06]. It is explained by different contact processes at the tool-chip and tool-workpiece interfaces. As a result, tool lives for the inserts with close relationships $U_{cs} = f(VB_B)$ but made of different tool materials may not be the same or close.

The foregoing analysis suggests that it is possible to choose the limiting tool wear, using the established correlation $U_{cs} = f(VB_B)$, and then calculate the limiting work, using this selected value. We can calculate tool life under a given cutting regime, using this limiting work, or the cutting speed for the desirable tool life by the following steps:

Calculation of tool life involves the following steps:

– Select the maximum allowed tool wear (VB_B).

– Calculate the limiting work, using correlation formulas similar to those shown in Table 2.10.

– Conduct a short cutting test for the chosen cutting regime to determine the cutting force P_z and the maximum cutting temperature.

– Calculate tool life as $T = U/(P_z v)$. If the accuracy of the cutting force determination is insufficient, then cutting energy can be determined using the chip compression ratio [AST 06] or direct measurements of the power of the driving motor.

– Compare the maximum cutting temperature obtained from the test and the limiting temperature for the selected tool material to ensure that the former is lower than the latter.

Calculation of the cutting speed for a given tool life involves the following steps:

– Choose the critical wear, VB_B, and the desired tool life, T.

– Calculate the limiting work, using correlation formulas similar to those shown in Table 2.10.

– Conduct a short cutting test establishing correlations $P_z = C_{P_z} v^{n_z}$ and $\theta = f(v)$.

– Calculate the cutting speed as $v = \sqrt[n_z+1]{U/(T C_{P_z})}$.

– Compare the maximum cutting temperature obtained from $\theta = f(v)$ and the limiting temperature for the selected tool material to ensure that the former is lower than the latter.

In the author's opinion, new reference books on cutting tools, cutting inserts, and tool materials issued and published by both the National Institutes for Standardization and the ISO should contain similar tables to help end users make a meaningful selection of the cutting tool and tool materials for particular applications.

2.7. References

[AST 06] ASTAKHOV V.P., *Tribology of Metal Cutting*, Elsevier, London, 2006.

[DEL 07] DELIU M., "A suggestion of new claims in tool quality standards", *Regent*, vol. 8, no. 3b, p. 460-462, 2007.

[BOB 02] BABU G.J., ANGELO J., CANTY A.J., CHAUBEY Y.P., "Application of Bernstein polynomials for smooth estimation of a distribution and density function", *Journal of Statistical Planning and Inference*, vol. 105, no. 2, p. 377-392, 2002.

[BAR 37] BARTLETT M.S., "Properties of sufficiency and statistical tests", *Proceedings of the Royal Society A*, vol. 160, p. 268-282, 1937.

[ELK 98] ELKINGTON J., *Cannibals with Forks: The Triple Bottom Line of 21st Century Business*, New Society Publishers, Stony Creek, CT, 1998.

[ELW 97] EL WARDANY T.I., ELBESTAWI M.A., "Prediction of tool failure rate in turning hardened steels", *International Journal of Advanced Manufacturing Technology*, vol. 13, no. 1, p. 1-16, 1997.

[HIT 79] HITOMI K., NAKAMURA N., INOUE S., "Reliability analysis of cutting tools", *Journal of Engineering for Industry – Transactions of the ASME*, vol. 101, p. 185-190, 1979.

[HUG 02] HUQ M.Z., CELIS J-P., "Expressing wear rate in sliding contacts based on dissipated energy", *Wear*, vol. 252, p. 375-383, 2002.

[KAT 74] KATSEV P.G., *Statistical Methods for Cutting Tools* (in Russian), Moscow, Mashinostroenie, 1974.

[KOM 02] KOMAROVSKY A.A., ASTAKHOV V.P., *Physics of Strength and Fracture Control: Fundamentals of the Adaptation of Engineering Materials and Structures*, Boca Raton, CRC Press, 2002.

[PAN 78] PANDID S.M., "Data dependant system approach to stochastic tool life and reliability", *ASME Journal of Engineering for Industry*, vol. 100, no. 1, p. 318-322, 1978.

[RAM 78-1] RAMALINGAM S., WATSON J.D., "Tool life distribution. Part 1: Single injury tool life model", *ASME Journal of Engineering for Industry*, vol. 99, no. 3, p. 519-522, 1978.

[RAM 78-2] RAMALINGAM S., "Tool life distribution. Part 2: Multiple-injury tool life model", *ASME Journal of Engineering for Industry*, vol. 99, no. 3, p. 523-531, 1978.

[RAM 78-3] RAMALINGAM S., PENG Y.L., WATSON J.D., "Tool life distribution. Part 3: Mechanism of single injury tool failure and tool life distribution in interrupted cutting", *ASME Journal of Engineering for Industry*, vol. 100, no.1, p. 193-200, 1978.

[RAM 78-4] RAMALINGAM S., WATSON J.D., "Tool life distribution. Part 4: Minor phases in work material and multiple-injury tool failure", *ASME Journal of Engineering for Industry*, vol. 100, no. 1, p. 201-209, 1978.

[ROC 09] ROCKWELL AUTOMATION, Manufacturing execution systems for sustainability. Extending the scope of MES to achieve energy efficiency and sustainability goals, Paper SUST-WP001A-EN-P, May 2009.

[UNI 87] *United Nations. Report of the World Commission on Environment and Development: Our Common Future*, General Assembly Resolution 42/187, 1987.

[SNE 89] SNEDECOR G.W., COCHRAN W., *Statistical Methods*, 8th ed., Iowa State University Press, Ames, 1989.

[WAN 01] WANG K.-S., LIN W.-S., HSU F.-S., "A new approach for determining the reliability of a cutting tool", *The International Journal of Advanced Manufacturing Technology*, vol. 17, no.10, p. 705-709, 2001.

[WAN 97] WANG K.-S., CHEN C.S., HUANG J.J., "Dynamic reliability behavior for sliding wear of carburized steel", *Reliability Engineering and System Safety*, vol. 58, p.31-41, 1997.

Chapter 3

Minimum Quantity Lubrication in Machining

This chapter deals with the importance of minimum quantity lubrication (MQL) and its benefits in machining. The investigations carried out by several researchers and their outcomes on the role of MQL in various machining processes are also discussed.

Furthermore, case studies involving MQL by artificial neural network (ANN) modeling and Taguchi optimization approaches are presented to illustrate the importance of MQL in turning.

3.1. Introduction

In this section, the importance of cutting fluids and problems related to cutting fluids, possible ways of tackling ecology problems such as dry cutting and MQL, the MQL technique, and its advantages and limitations are discussed.

Chapter written by Vinayak N. GAITONDE, Ramesh S. KARNIK and J. Paulo DAVIM.

3.1.1. *Cutting fluids and problems related to cutting fluids*

Cutting fluids are used in machining processes to improve the characteristics of tribological processes, which are present on the contact surfaces between the workpiece and the cutting tool. The cutting fluid increases the surface finish of a machined component and improves the tool life. The cutting fluid acts both as a lubricant and as a coolant during the cutting operation. Moreover, the fluid protects the finished surface from scratches.

An improper disposal of cutting fluids causes environmental problems such as water and soil pollution [WAG 98]. Therefore, handling and disposal of cutting fluids must obey rigid rules of environmental protection. In machine shop, the operators may suffer from adverse effects of cutting fluids such as skin and breathing problems [SOK 01]. It was reported that because of toxicity, there might be health problems such as dermatitis, respiratory and digestive system disorders, and even cancer among operators who are exposed to cutting fluids [THE 09]. The total production cost also increases as a result of procurement, storage, maintenance, and disposal of cutting fluids. As reported by Byrne and Scholta [BYR 93] and Klocke and Eisennblatter [KLO 97], the costs related to cutting fluids are frequently higher than those of cutting tools. Furthermore, a survey by the European automotive industry revealed that the expense of cooling lubricant makes up approximately 20% of the total production cost [BRO 98].

The minimization of cutting fluids leads to economical benefits such as saving of lubrication costs and workpiece/tool/machine cleaning cycle time. Hence, elimination of the use of excessive cutting fluids, if possible, can be a significant economic incentive [DHA 07]. Because of high costs associated with the use of cutting fluids and projected escalating costs when the stricter environmental laws are enforced, the choice of avoiding cutting fluids seems to be obvious [DHA 07]. On the other hand, despite several attempts to completely eliminate the cutting fluids, it may not be possible because cooling is still necessary for economic service life of cutting tools and required surface qualities. This is particularly true when high tolerance and high-dimensional and shape exactness are required, or when the machining of critical and hard materials are involved [SIL 07]. Hence, there has been increasing attention on the proper selection of cutting fluids from the viewpoint of cost, ecology, and environmental issues in the modern production industries [SRE 00].

3.1.2. *Dry cutting and its limitations*

Because of several negative effects, continuous efforts have been made to minimize or even completely eliminate the use of cutting fluids [SOK 01]. Dry cutting or cutting with no fluid is preferred in the field of environmentally friendly manufacturing [CAN 05] and is found to be effective in machining of hardened steel parts, which show a very high specific cutting energy. Furthermore, dry machining lowers the required cutting force and power on machine tool parts as a result of increased cutting temperature. However, dry machining is less effective for achieving good surface finish, economical tool life, and higher machining efficiency. Therefore, the permissible feed and depth of cut have to be restricted and under these conditions, the MQL finds a possible and an alternative solution for hard turning for minimizing tool wear while maintaining the cutting forces/power at reasonable levels, provided that the MQL parameters can be strategically tuned.

3.1.3. *MQL and its performance in machining*

The MQL in machining, sometimes known as near-dry lubrication [KLO 97], is an alternative to a completely dry or flood lubricating system [DIN 03], which is considered as one of the solutions for reducing the amount of lubricant that addresses the environmental, economical, and mechanical process performance issues [HEI 06]. The MQL refers to the use of cutting fluids in a small amount, flow rate typically in the range of 50 to 500 mL/h, which is about three to four orders of magnitude lower than the amount commonly used in flood cooling condition [DHA 06a]. Because of favorable properties such as advanced thermal stability and lubricating capability, the MQL has the edge over the conventional cutting fluids. One more advantage of MQL is that it combines the functionality of cooling with an extremely low consumption of fluids (<80 mL/h), which, in turn, reduce the tool's friction and prevent the adherence of materials [SIL 07].

The MQL technique consists of atomizing a very small quantity of lubricant in an airflow directed toward the cutting zone [AVI 01, DIN 03, KLO 97, WEI 04]. The aerosol (often referred to as mist) can be sprayed by either an external supply system through one or more nozzles or an internal supply system through channels built inside the cutting tool. The former is useful in the application of various processes such as turning, milling, sawing, drilling, reaming, and tapping with low values of length to diameter ratio (< 3), whereas the latter is beneficial for higher length to diameter ratio values and essential in deep-hole drilling operations [BRU 06]. In MQL, the cooling ability depends only on the airflow and hence it must be difficult to completely replace the flood coolant techniques.

The MQL with water droplets, called OoW (oil film on water droplet), has a large cooling ability because water droplets play the role of an oil carrier and also easily evaporate from the cutting tool and work surfaces because of their size. These also chill the surfaces by their sensible and latent heat [ITO 06]. This cooling ability is important not only for the dimensional accuracy but also for the tribological phenomenon, such as adhesion, between the cutting tool and the work surface. In addition, the water droplets ensure that lubricant coating deposits and spreads over the tool and work surfaces because of the inertia of droplets.

The MQL more often uses straight oils rather than vegetable oils, which are chosen for their high lubricity as compared with mineral oils. In recent times, the most common minimum quantity lubricants are polyglycols, ester oils, and fatty alcohols. The polyglycols and ester oils have good lubrication capability and low evaporation tendency, whereas fatty alcohols show an improved cooling capability because of their high evaporation tendency. In addition, the use of biodegradable oils that do not contain harmful elements and compounds such as extreme pressure additives is also recommended.

3.1.4. *Limitations of MQL*

The limitation of MQL lies in its inability to cool the cutting surface. If MQL is applied especially for a grinding operation, which requires a strong cooling action, the MQL does not serve its purpose. Hence, it is essential to define the conditions that allow the MQL to be applied with real advantages. Furthermore, the MQL has poor ability in the removal of the cutting edge from the cutting area. The suction of floating oil mist inside the body is harmful for health. The oil mist adheres to the inside of the machine tool or machine shop floor, and it may cause slipping accidents [AOY 08]. It was reported that the floating mist generated in the MQL in machining might cause lung disease if oil mist less than 10 μm in diameter enters deep into the lungs. Therefore, it is important to spray a proper amount of oil mist to the cutting point [OBI 08].

Because of medium cooling ability, the MQL is favorable for intermittent cutting with thermal shock [LIA 07] but not for the machining of difficult-to-machine materials with intense heat generation. Nevertheless, there is a demand to apply MQL for the machining of difficult-to-machine materials from an environmental viewpoint [OBI 09]. As compared with conventional technique, the MQL involves additional cost to pressurize the air and support resources required to overcome the technological restrictions of MQL [SIL 05].

3.2. The state-of-the-art research for MQL in machining

A survey of the literature reveals that there is an increasing trend in the application of MQL in various machining processes. In this section, a review of some such research efforts performed on MQL in various machining processes is presented. The experimental investigations and their outcomes on the role of MQL are also discussed.

3.2.1. *Experimental studies on MQL in drilling*

Braga *et al.* [BRA 02] applied the MQL in drilling of aluminum-silicon alloys with 7% silicon (SAE 323) with uncoated and diamond-coated K10 carbide tools. It was observed that the quality of holes obtained under the MQL condition was better than that obtained with the flood lubricant system. The investigations carried out by Kelly and Cotterell [KEL 02] during drilling of aluminum-magnesium alloy revealed that the MQL could be preferred for higher cutting speeds and feed rates. An experimental investigation carried out by Hanyu *et al.* [HAN 03] on dry and semi-dry drilling of 12% silicon-aluminum alloy (JISADC12) with finely crystallized smooth surface diamond coatings showed that a smooth surface diamond coating presents longer durability of drills than conventional rough surface diamond coating under both dry and semi-dry cutting conditions.

Davim et al. [DAV 06] presented a comparative study on drilling experiments of aluminum alloys (AA 1050) with K10 carbide tools under dry, MQL, and flooded lubricated conditions. The effects of cutting speed and feed rate on cutting power, specific cutting force, and surface roughness were studied under different lubricated conditions. The authors reported that with a proper selection of machining parameters, it is possible to obtain a performance similar to that of a flood coolant by MQL. Mendes et al. [MEN 06] discussed the performance of cutting fluids during drilling of AA 1050-O aluminum alloys. The results indicated that an increased in-flow rate of mist led to lower feed forces without affecting the surface finish significantly.

The effect of MQL in drilling of Al-6% Si (319 Al) aluminum alloy was examined by Bhowmick and Alpas [BHO 08] with both high-speed steel (HSS) and diamond-like carbon (DLC)-coated drills and the results were thus compared with a conventional flooded coolant. The smallest built-up edge formation on the cutting edge of the drill and aluminum adhesion on the drill flutes were observed for DLC-coated drills by MQL.

A study on the prediction of machining performances such as surface roughness, cutting power, and specific cutting force with MQL in drilling of aluminum (AA 1050) by Fuzzy logic rules was carried out by Nandi and Davim [NAN 09]. A significant improvement in the machinability characteristics was observed by MQL. Heinemann et al. [HEI 06] presented an experimental investigative study on plain carbon steel (0.45% C) in deep-hole drilling under the MQL condition with small-diameter HSS twist drills, and it was found that the discontinuous supply of MQL could reduce tool life as compared with continuous supply of MQL. Zeilmann and Weingaertner [ZEI 06] summarized temperature analysis in the drilling of titanium alloy (Ti6Al4V) with different types of coated K10 carbide drills under MQL conditions. In their investigations, drilling with MQL applied internally through the drill was better than the lubricant applied with an external nozzle. Tasdelen *et al.* [TAS 08a] investigated the effect of oil droplets and air in the aerosol at MQL cutting with different oil amounts, dry compressed air, and emulsion in short-hole drilling. The drilling tests with indexable inserts confirmed that the MQL and compressed air usage have resulted in less wear on the central and peripheral inserts than drilling with emulsion.

3.2.2. *Experimental studies on MQL in milling*

Rahman *et al.* [RAH 01] conducted experiments on the milling of ASSAB 718HH steel workpieces with carbide end mills under MQL and flood coolant systems, and the comparative effectiveness was measured in terms of cutting force, tool wear, surface roughness, and chip shape. It was concluded that the MQL system could provide an alternative to flood coolant system from the viewpoint of both economical and environmental issues. Kishawy *et al.* [KIS 05] studied the effects of flood coolant, dry cutting, and MQL techniques on tool wear, surface roughness, and cutting forces in high-speed milling of aluminum alloy (A356) and found that the MQL technique is a better alternative to flood coolant application. The experimental research carried out by Sun *et al.* [SUN 06] on titanium alloy milling under dry cutting, flood cooling, and MQL techniques showed that the MQL machining could improve tool life and reduce cutting force as a result of better lubrication and cooling effect.

Lopez de Lacalle *et al.* [LOP 06] carried out both experimental and numerical investigations on high-speed milling of wrought aluminum. The machining performance under MQL was assessed through the computational fluid dynamics simulation and experimental evidences. Kang *et al.* [KAN 08] studied the effect of MQL in high-speed end milling of AISI D2 cold-worked die steel on tool wear with various types of coated carbide tools. It was observed that cutting under flood coolant condition results in shortest tool life because of severe thermal cracks, whereas the use of MQL leads to the best performance. A new lean lubrication system for a near-dry machining process called "direct oil drop supply system (DOS)" was developed by Aoyama et al. [AOY 08] and the performance was evaluated by milling processes. It was found that the proposed DOS has the possibility of reduction in the total amount of oil consumption by decreasing the diameter of oil drops. Iqbal et al. [IQB 08] modeled the effects of cutting parameters in MQL-employed finish hard-milling process with coated carbide ball-nose end mills. In their studies, the D-optimal technique was applied to investigate the influence of workpiece material, inclination angle of workpiece, rotational speed of tool, and radial depth of cut on tool life and surface roughness. The machinability study carried out by Ahmed et al. [AHM 09] on the milling of titanium alloy (Ti6Al4V) with a physical vapor deposition–coated cemented carbide end-mill cutter under dry and MQL conditions concluded that the MQL is more effective when a worn-out tool is applied. Furthermore, it was also observed that more contact area between cutting tool and workpiece provides better lubrication effect. Thepsonthi et al. [THE 09] introduced a pulsed-jet application technique in high-speed milling of hardened steel with carbide mills. The performance of machining was compared with dry machining and machining with flood application. The results clearly indicated that the performance of pulsed-jet application was more than that of dry cutting and flood application, based on cutting forces, tool wear, tool life, and surface finish especially when machining with high cutting velocity.

3.2.3. *Experimental studies on MQL in turning*

Dhar et al. have to be credited for their detailed study on MQL in turning. They [DHA 06a] used the MQL technique in turning of AISI 1040 steel and concluded that machining with a mixture of air and soluble oil is better than the conventional flood coolant system. Dhar et al. [DHA 07] carried out further investigations on AISI 1040 steel and compared the mechanical performance of MQL with dry lubrication. It was revealed that MQL machining is better than dry machining on the basis of experimental measurement of cutting temperature, chip reduction coefficient, cutting forces, tool wear, surface finish, and dimensional deviation. Dhar et al. [DHA 06b] also performed experiments to analyze the effect of MQL on tool wear and surface roughness in turning of AISI 4340 steel workpieces at industrial speed-feed combination with uncoated carbide inserts. The experimental results showed that tool wear rate and surface roughness could be reduced by MQL through reduction in cutting zone temperature and favorable change in chip-tool and work-tool interaction.

Attanasio et al. [ATT 06] observed the effect of MQL on tool wear in 100Cr6 normalized steel turning. The results obtained from the turning tests showed that when MQL is applied to the tool rake, tool life is generally no different from that under dry conditions but MQL applied to the tool flank can increase tool life. The effects and mechanisms in MQL of aluminum-silicon alloys (Al-Si5) in an intermittent turning process with sintered diamond and K10 carbide tools was studied by Itoigawa et al. [ITO 06]. The experiments were conducted with various lubrication methods such as MQL and MQL with water to elucidate the boundary film behavior on rake face. Mendes et al. [MEN 06] conducted experiments on turning of 6262-T6 aluminum alloy with plain carbide inserts and the effect of additives on the performance of cutting fluid in terms of cutting force and surface roughness was investigated.

Davim et al. [DAV 07] studied the influence of cutting speed and feed rate on machinability characteristics in turning of brass under different conditions of lubricant environments. The authors reported that the MQL-type lubrication could successfully replace the flood lubrication in machining of brass. Kamata and Obikawa [KAM 07] used MQL to observe the effects of higher values of cutting speed on finish turning of a nickel-base super alloy (Inconel 718) with different types of coated carbide tools. The study confirmed that there was a certain spraying pressure, which provides the longest tool life.

The effect of oil and air in MQL on contact length was evaluated by Tasdelen *et al.* [TAS 08b] in intermittent turning of 100Cr work material with uncoated and TiN-coated inserts. They illustrated that the MQL is a suitable method for short engagement time in machining. Khan *et al.* [KHA 09] explained the effects of MQL with vegetable oil-based cutting fluid on the turning performance of low alloy steel (AISI 9310) with carbide inserts based on chip-tool interface temperature, chip formation mode, tool wear and surface roughness. The performance was found to be greater than those of dry and wet machining.

3.2.4. *Experimental studies on MQL in other machining processes*

In addition to these drilling, milling, and turning processes, a few authors have also reported the MQL in other machining processes such as grinding, reaming, tapping, and grooving.

The experimental tests carried out by Silva *et al.* [SIL 05] on plunge cylindrical grinding of annealed ABNT 4340 steel indicated that the surface roughness and diametral wear values were substantially reduced with the use of MQL. The performance of MQL in the grinding process was evaluated by Silva *et al.* [SIL 07] on surface roughness, residual stress, micro-hardness, and micro-structure. The results presented under MQL are expected to lead to technological and ecological gains in the grinding process.

Lugscheider *et al.* [LUG 97] used MQL in reaming of gray cast iron (GG25) and aluminum alloy (Al-Si12) with coated carbide tools. The significant reduction in tool wear and improvement in surface quality of the holes have been observed. Sokovic and Mijanovic [SOK 01] conducted experiments on tapping, which is a multi-point cutting operation and considered to be one of the most difficult operations in metal removal. The results of machinability tests on C 45 E4 steel and AlMgSiPbBi aluminum alloy confirmed the suitability of MQL. The high-speed grooving of a carbon steel with P35-coated tool under MQL, dry, and wet conditions was carried out by Obikawa *et al.* [OBI 06] to investigate the performance of MQL. The MQL reduced both the corner and flank wears more effectively than a solution-type cutting fluid.

3.3. Case studies on MQL in machining

The majority of the published works highlighted several benefits of MQL, including the role and effectiveness of MQL in various machining processes over dry and wet machining conditions. No systematic work has been reported in the previous studies to analyze the effects of MQL and cutting conditions on the performance characteristics in any of the machining processes. Moreover, optimization of performance characteristics for the determination of optimal MQL and cutting conditions has not been addressed so far.

The surface roughness is a widely used product quality index and, in most cases, a technical requirement for mechanical products. For the functional behavior of the part, achieving a desired surface quality is of greater concern [BEN 03]. The surface roughness plays an important role because it influences fatigue strength, corrosion resistance, coefficient of friction, and wear rate on machined components. For achieving a better surface finish on the machined part with minimum specific cutting force, either the cutting conditions should be carefully selected or a new tool material should be replaced with lower coefficient of friction and high heat resistance. Furthermore, the amount of MQL plays a significant role in the surface finish, which needs to be investigated. In this regard, two case studies are presented in this section to meet these requirements. The first one is aimed at studying the effects of MQL and cutting conditions on machinability aspects in turning of brass. On the other hand, the objective of second case study is to determine the optimal quantity of lubricant with appropriate cutting conditions for achieving a better machinability in turning of brass.

3.3.1. *Case study 1: analysis of the effect of MQL on machinability of brass during turning – ANN modeling approach*

3.3.1.1. *Relevance of machinabilty modeling*

The effects of cutting speed, feed rate, and MQL on machinability aspects such as specific cutting force and surface roughness in brass turning with a K10 carbide tool has been analyzed in this section. The model development by response surface methodology (RSM) is a convenient method, which requires minimum number of experiments to be performed and thus reducing the cost and time [MON 01]. However, the RSM-based mathematical models are restricted to only a small range of input parameters and hence not suitable for highly complex and nonlinear processes. On the other hand, the development of higher-order RSM models requires an increased number of experiments to be conducted and hence is costlier. This poses a limitation on the use of RSM models for highly nonlinear process. These constraints have led to the development of a model based on ANN.

The ANN is a fast, efficient, accurate, and cost-effective process modeling tool in which the biological neurons are represented by a mathematical model [SCH 97]. The ability of ANN to capture any complex input-output relationship from the limited data set is valuable especially in the machining processes, where a huge experimental dataset for the process modeling is difficult and expensive to obtain.

3.3.1.2. *Experimental plan*

A multi-layered, feed-forward ANN) trained by error backpropagation training algorithm (EBPTA) [SCH 97] has been used to capture the relationship between MQL (Q), cutting speed (v), and feed rate (f) on specific cutting force (K_s) and surface roughness (R_a) [GAI 10]. The input-output data required for the development of ANN model has been obtained through full factorial design (FFD) of experiments. The experimental factors and their levels are illustrated in Table 3.1. The results presented by Gaitonde *et al.* [GAI 10] were obtained by turning of Cu-Zn39-Pb3 brass work material with hardness of 66 HRB.

Factor	Notation	Unit	Levels			
			1	2	3	4
MQL	Q	mL/h	50	100	200	–
Cutting speed	v	m/min	100	200	400	–
Feed rate	f	mm/rev	0.05	0.10	0.15	0.20

Table 3.1. *Experimental factors and their levels for ANN modeling (adapted from Gaitonde et al. 2010, with permission from Interscience Enterprises Ltd. [GAI 10])*

3.3.1.3. *Experimental procedure*

The Kingsbury MHP 50 CNC lathe operating on 18-kW spindle power with a maximum spindle speed of 4,500 rpm has been employed for the turning tests. K10 carbide inserts of TCGX 16 T3 08-Al H10 (Sandvik) were used throughout the machining tests. The tool geometry of carbide inserts is as follows: rake angle: 15°; clearance angle: 7°; major edge cutting angle: 91°; cutting edge inclination angle: 0°; and corner radius: 0.8 mm. The STGCL2020K16-type tool holder was used throughout the experimental study. The experiments were performed as per FFD under different combinations of MQL (lubricated with emulsion oil, Microtrend 231L), cutting speed, and feed rate. The depth of cut of 2 mm was kept constant throughout experimental investigation.

The acquisition of the cutting force (F_c) was performed by a Kistler® (model 9121) piezoelectric dynamometer. The cutting force values were continuously monitored and recorded throughout the study with a three-channel (model 5019) charge amplifier with data acquisition. The specific cutting force (K_s) is determined as follows:

$$K_s = \frac{F_c}{f \cdot d} \qquad [3.1]$$

where d is the depth of cut. The surface roughness (R_a) on turned components was measured with a Hommelwerke® T1000 profilometer with a cut-off distance of 0.8 mm in accordance with ISO/DIS 4287/1E. The layout plan and the experimental results for 36 sets of factor combinations reported by Gaitonde et al. [GAI 10] are summarized in Table 3.2.

3.3.1.4. ANN training and testing

ANN training was performed using first 27 input-output patterns of the experimental plan as shown in Table 3.2, whereas the last 6 data sets of Table 3.2 were used for ANN testing. Before ANN training, all inputs and desired or target outputs are normalized by:

$$X_{norm} = \frac{2(X - X_{min})}{(X_{max} - X_{min})} - 1 \qquad [3.2]$$

where X_{min} and X_{max} are the minimum and maximum values in the vector pattern for X.

Trial no.	Process parameter settings			Responses	
	Q (mL/h)	v (m/min)	f (mm/rev)	K_s (MPa)	R_a (μm)
1	50	100	0.1	323.5	0.54
2	50	100	0.15	317	1.07
3	50	100	0.2	274.6	1.93
4	50	200	0.05	364.8	0.28
5	50	200	0.1	323	0.5
6	50	200	0.15	299.1	1.1
7	50	400	0.05	397.2	0.28
8	50	400	0.1	337.6	0.49
9	50	400	0.2	283.3	1.68
10	100	100	0.05	359.1	0.28
11	100	100	0.1	307.4	0.44
12	100	100	0.15	282.9	0.88
13	100	100	0.2	262.5	1.6
14	100	200	0.1	316	0.45
15	100	200	0.2	267.9	1.59
16	100	400	0.1	314.2	0.6
17	100	400	0.15	278.1	1.02
18	100	400	0.2	254.4	1.69
19	200	100	0.05	350.6	0.24
20	200	100	0.1	301.3	0.37
21	200	100	0.2	286	1.5
22	200	200	0.05	343	0.24
23	200	200	0.1	299	0.42
24	200	200	0.15	280.9	0.97
25	200	400	0.1	317.9	0.42
26	200	400	0.15	273.1	0.95
27	200	400	0.2	265	1.61
28	50	100	0.05	375.1	0.3
29	50	400	0.15	301.5	1.1
30	100	200	0.15	286.1	0.9
31	100	400	0.05	374.9	0.32
32	200	100	0.15	304.3	0.78
33	200	200	0.2	256.3	1.61
34	100	200	0.05	360.2	0.29
35	50	200	0.2	281.8	1.73
36	200	400	0.05	369.2	0.26

Table 3.2. *Experimental layout plan and responses as per FFD for ANN modeling (adapted from Gaitonde et al. 2010, with permission from Interscience Enterprises Ltd. [GAI 10])*

This normalization maps all the inputs and target outputs in the range [−1:1]. The ANN designed for the present study takes MQL, cutting speed, and feed rate as the input parameters and specific cutting force and surface roughness as the output parameters. Two separate ANN models of specific cutting force and surface roughness were developed The ANN architecture selected for the specific cutting force model is 3-7-1, whereas it is 3-5-7-1 for surface roughness model. The details of the ANN training, testing, and the learning factors employed are given in Gaitonde *et al.* [GAI 10].

3.3.1.5. *Parametric analysis on machinability aspects in brass turning*

The developed ANN models of specific cutting force and surface roughness are used to analyze the interaction effects of process parameters by generating three-dimensional (3D) surface plots. In this, the variation in two parameters at a time are considered whereas the third parameter is kept constant at its center level [GAI 10].

Figures 3.1, 3.2, and 3.3, as obtained from an ANN model of specific cutting force [GAI 10], illustrate the interaction effects of process parameters on specific cutting force. It is observed from Figure 3.1 that the minimum specific cutting force exists at high MQL with the medium range of cutting speed. On the other hand, the higher specific cutting force results at higher cutting speed with low MQL. As seen from Figure 3.2, the specific cutting force is minimum at high feed rate with low to medium MQL. The specific cutting force is highly sensitive to feed rate variations for all values of MQL. The specific cutting force is also minimum at higher feed rate, with a medium to high range of cutting speed as depicted in Figure 3.3.

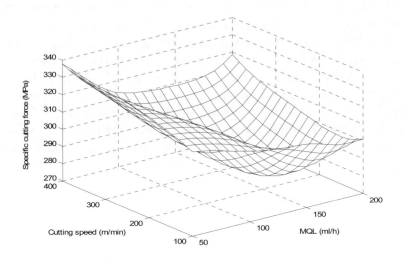

Figure 3.1. *Interaction effect of MQL and cutting speed on specific cutting force for feed rate of 0.1 mm/rev (adapted from Gaitonde et al. 2010, with permission from Interscience Enterprises Ltd. [GAI 10])*

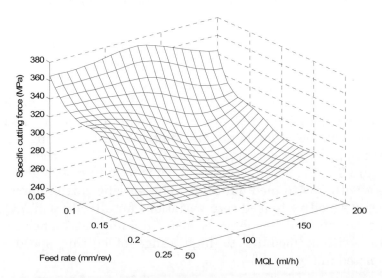

Figure 3.2. *Interaction effect of MQL and feed rate on specific cutting force for cutting speed of 200 m/min (adapted from Gaitonde et al. 2010, with permission from Interscience Enterprises Ltd. [GAI 10])*

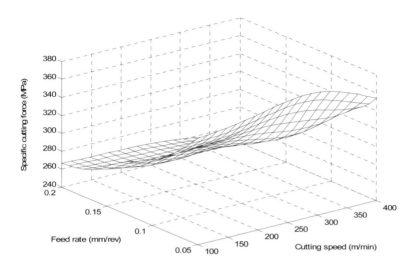

Figure 3.3. *Interaction effect of cutting speed and feed rate on specific cutting force for MQL of 100 mL/h (adapted from Gaitonde et al. 2010, with permission from Interscience Enterprises Ltd. [GAI 10])*

The interaction effects of process parameters on surface roughness as reported by Gaitonde *et al.* [GAI 10] are presented in Figures 3.4, 3.5, and 3.6. The minimum surface roughness results in high MQL with low values of cutting speed and the surface roughness is highly sensitive to MQL variations at higher values of cutting speed as shown in Figure 3.4.

As can be seen from Figure 3.5, the surface roughness sharply increases with increase in feed rate irrespective of MQL and the variation in surface roughness is minimum with variation in MQL at all values of feed rate. It was observed from Figure 3.6 that the surface roughness increases with an increase in feed rate irrespective of the cutting speed.

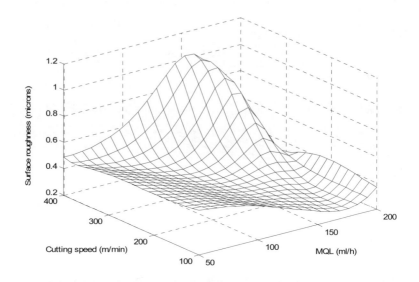

Figure 3.4. *Interaction effect of MQL and cutting speed on surface roughness for feed rate of 0.1 mm/rev (adapted from Gaitonde et al. 2010, with permission from Interscience Enterprises Ltd. [GAI 10])*

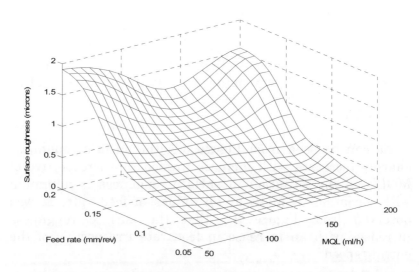

Figure 3.5. *Interaction effect of MQL and feed rate on surface roughness for cutting speed of 200 m/min (adapted from Gaitonde et al. 2010, with permission from Interscience Enterprises Ltd. [GAI 10])*

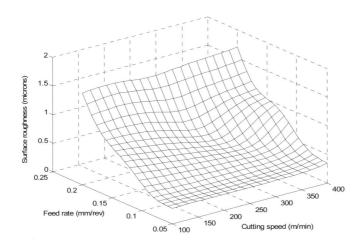

Figure 3.6. *Interaction effect of cutting speed and feed rate on surface roughness for MQL of 100 mL/h (adapted from Gaitonde et al. 2010, with permission from Interscience Enterprises Ltd. [GAI 10])*

From this discussion, it is clear that a combination of medium to high MQL, higher feed rate, and cutting speed in medium range is essential for minimizing the specific cutting force. On the other hand, a combination of lower values of feed rate and cutting speed with high MQL is necessary for minimizing the surface roughness. Therefore, the combination of MQL, cutting speed, and feed rate required for minimizing the specific cutting force may not be the same for minimizing the surface roughness. This requires simultaneous optimization of specific cutting force and surface roughness using an efficient multi-response optimization tool.

3.3.2. *Case study 2: selection of optimal MQL on machinability of brass during turning – Taguchi approach*

The study presented here describes the selection of optimal MQL and the most appropriate cutting speed and feed rate during turning of brass with a K10 carbide tool

[GAI 08]. The Taguchi technique with a utility concept, a multi-response optimization method, has been employed for simultaneous minimization of surface roughness and specific cutting force.

3.3.2.1. *Taguchi design and utility concept*

The Taguchi design is based on matrix experiments. Taguchi developed the procedures that apply orthogonal arrays (OA) of designed experiments to obtain the best model with reduced number of experiments, thus minimizing time and cost of experimentation [PHA 89, ROS 96]. Taguchi suggested signal to noise (S/N) ratio as the objective function for matrix experiments, which is used to measure the quality characteristics. The S/N ratio also indicates the degree of predictable performance in the presence of noise factors. Taguchi classified the S/N ratio into smaller-the-better type, larger-the-better type, and nominal-the-best type, which are based on the type of objective function [PHA 89, ROS 96]. The optimal level for a parameter is the level that results in the highest value of S/N ratio in the experimental region.

The original Taguchi technique is designed for optimizing a single performance characteristic. However, most of the products/processes have several performance/quality characteristics and hence there is a need to obtain a single optimal process parameter setting. Several modifications are suggested in the literature to the original Taguchi optimization technique for multi-performance characteristics optimization. This study presents the application of the Taguchi method and the utility concept for multiple quality characteristics, which employs the weighting factors to each of the S/N ratio of the responses to obtain a multi-response S/N ratio for each trial of an orthogonal array [KUM 00].

3.3.2.2. *Experimental details*

The identified parameters, namely, quantity of lubricant (Q), cutting speed (v), and feed rate (f) and their selected levels during turning of brass as reported in Gaitonde *et al.* [GAI 08] are given in Table 3.3. As in the previous case study, the workpiece material, cutting tool, and the same experimental setup were employed in the current investigation. According to the Taguchi design, for three levels and three factors, nine experiments are required and hence L_9 OA was selected [PHA 89, ROS 96]. The OA and the experimental results as reported in Gaitonde *et al.* [GAI 08] are given in Table 3.4.

Factor	Code	Unit	Levels		
			1	2	3
Quantity of lubricant Q	A	mL/h	50	100	200
Cutting speed v	B	m/min	100	200	400
Feed rate f	C	mm/rev	0.05	0.10	0.15

Table 3.3. *Process parameters and their levels for the Taguchi optimization (adapted from Gaitonde et al. 2008, with permission from Elsevier [GAI 08])*

Trial no.	Levels of input process parameters			Responses		Calculated S/N ratio (dB)		
	A	B	C	R_a (µm)	K_s (MPa)	η_1	η_2	η
1	1	1	1	0.3	375.1	10.4576	−51.4829	−20.5127
2	1	2	2	0.5	323	6.0206	−50.1841	−22.0817
3	1	3	3	1.1	301.5	−0.8279	−49.5857	−25.2068
4	2	1	2	0.44	307.4	7.1309	−49.7541	−21.3116
5	2	2	3	0.9	286.1	0.9151	−49.1304	−24.1076
6	2	3	1	0.32	374.9	9.8970	−51.4783	−20.7907
7	3	1	3	0.78	304.3	2.1581	−49.6660	−23.7540
8	3	2	1	0.24	343	12.3958	−50.7059	−19.1551
9	3	3	2	0.42	317.9	7.5350	−50.0458	−21.2554

Table 3.4. *L_9 Orthogonal array, responses and the calculated values of S/N ratio of multi-performance characteristic for the Taguchi optimization (adapted from Gaitonde et al. 2008, with permission from Elsevier [GAI 08])*

3.3.2.3. *Data analysis*

The S/N ratio for the selected performance characteristics (smaller-the-better type) is given by [GAI 08]:

$$\eta_1 = -10\log_{10}[R_a^2] \quad [3.3]$$

$$\eta_2 = -10\log_{10}[K_s^2] \quad [3.4]$$

In the utility concept, the multi-response S/N ratio for each trial in an OA is determined by [KUM 00]:

$$\eta = w_1\eta_1 + w_2\eta_2 \quad [3.5]$$

where w_1 and w_2 are the weighting factors associated with S/N ratio for each of the characteristics R_a and K_s, respectively. The weighing function values for w_1 and w_2 are selected on the basis of priority so as to satisfy the condition $(w_1 + w_2) = 1.0$. A weighting factor of 0.5 for each of the characteristics is selected, which gives an equal importance to R_a and K_s for the simultaneous minimization. The calculated values of S/N ratio for each characteristic and multi-response S/N ratio (η) for each trial in OA as reported in Gaitonde *et al.* [GAI 08] are summarized in Table 3.4.

The analysis of means, which gives the optimal process parameter combination, and the results of analysis of variance (ANOVA), which summarizes the percentage contribution of each factor, are given in Figure 3.7 and Table 3.5, respectively. The optimum factor level combinations as reported by Gaitonde *et al.* [GAI 08] are quantity of lubricant 200 mL/h; cutting speed 200 m/min, and feed rate 0.05 mm/rev. The details of the two-factor interaction effects of process parameters and the prediction and verification of the quality characteristic using the optimal level of the design parameters are given in Gaitonde *et al.* [GAI 08].

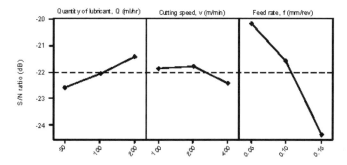

Figure 3.7. *Response plot for multi-performance characteristics for the Taguchi optimization (adapted from Gaitonde et al. 2008, with permission from Elsevier [GAI 08])*

Factor	Degrees of freedom	Sum of squares	Mean square	% contribution
A	2	2.2158	1.1079	7.24
B	2	0.7224	0.3612	2.36
C	2	27.4957	13.7478	89.85
Error	2	0.1681	0.0840	0.55
Total	8	30.6020	3.8252	100

Table 3.5. *Results of ANOVA for multi-performance characteristics for Taguchi optimization (adapted from Gaitonde et al. 2008, with permission from Elsevier [GAI 08])*

3.3.2.4. *Discussion on Taguchi optimization results*

From the Taguchi optimization results [GAI 08], it was observed that high MQL is required for minimizing both surface roughness and specific cutting force. This is because high MQL improves surface finish depending on work-tool material through controlling the deterioration of auxiliary cutting edge of abrasion, chipping, and built-up edge formation. Furthermore, at high MQL, the specific cutting force decreases due to reduction in cutting temperature, especially at the main cutting edge where built-up edge formation is more predominant.

As reported in Gaitonde *et al.* [GAI 08], high MQL with low to medium cutting speed is also essential for minimizing both surface roughness and specific cutting force. Furthermore, it was clear from this investigation that the feed rate required is low for minimizing both the performance characteristics. This is because under these conditions, there will be reduction in adhesion of work material and consequently the friction forces, which, in turn, reduces the specific cutting force. On the other hand, increase in the feed rate increases chatter, providing an incomplete machining at a faster traverse rate that results in higher surface roughness.

3.4. Summary

This chapter addressed the importance of MQL in the machining process. A detailed review of the state-of-the-art MQL in various machining processes reported in the literature such as drilling, milling, turning, grinding, reaming, tapping, and grooving is presented.

Two case studies involving MQL investigations on machinability aspects such as specific cutting force and surface roughness during turning of brass with a K10 carbide tool were presented. The first investigation is on the ANN modeling approach to study the influence of MQL and cutting conditions on machinability. In this study, two separate ANN models for specific cutting force and surface roughness were developed. The two-factor interaction effects on machinability were studied by generating 3D surface plots, which demonstrated the importance of MQL. Taguchi-based MQL optimization using utility concept is detailed in the second case study. Here, the optimum values of MQL, cutting speed, and feed rate were determined for simultaneous minimization of specific cutting force and surface roughness. The investigative case studies revealed that high MQL values in the selected range (150–200 mL/h) is appropriate to achieve better machinability during turning of brass material with a carbide tool.

3.5. Acknowledgments

The authors would like to thank Elsevier and Interscience Enterprises Ltd. for granting permission for the re-use of the published materials.

3.6. References

[AHM 09] AHMAD YASIR M.S., CHE HASSAN C.H., JAHARAH A.G., NAGI H.E., YANUAR B., GUSRI A.I., "Machinability of Ti-6Al-4V under dry and near dry condition using carbide tools", *The Open Industrial and Manufacturing Journal*, vol. 2, p. 1-9, 2009.

[AOY 08] AOYAMA T., KAKINUMA Y., YAMASHITA M., AOKI M., "Development of a new lean lubrication system for near dry machining process", *CIRP Annals – Manufacturing Technology*, vol. 57, p. 125-128, 2008.

[ATT 06] ATTANASIO A., GELFI M., GIARDINI C., REMINO C., "Minimal quantity lubrication in turning: effect on tool wear", *Wear*, vol. 260, p. 333-338, 2006.

[AVI 01] AVILA R.F., ABRAO A.M., "The effect of cutting fluids on the machining of hardened AISI 4340 steel", *Journal of Materials Processing Technology*, vol. 119, p. 21-26, 2001.

[BEN 03] BENARDOS P.G., VOSNIAKOS G.C., "Predicting surface roughness in machining: a review", *International Journal of Machine Tools and Manufacture*, vol. 43, p. 833-844, 2003.

[BHO 08] BHOWMICK S., ALPAS T., "Minimum quantity lubrication drilling of aluminum-silicon alloys in water using diamond-like carbon coated drills", *International Journal of Machine Tools and Manufacture*, vol. 48, p. 1429-1443, 2008.

[BRA 02] BRAGA D.U., DINIZ A.E., MIRANDA G.W.A., COPPINI N.L., "Using a minimum quantity of lubricant (MQL) and a diamond coated tool in the drilling of aluminum-silicon alloys", *Journal of Materials Processing Technology*, vol. 122, p. 127-138, 2002.

[BRO 98] BROCKHOFF T., WALTER A. "Fluid minimization in cutting and grinding", *Abrasives*, p. 38-42, October 1998.

[BRU 06] BRUNI C., FORCELLESE A., GABRIELLI F., SIMONCINI M., "Effect of the lubrication-cooling technique, insert technology and machine bed material on the work part surface finish and tool wear in finish turning of AISI 420B", *International Journal of Machine Tools and Manufacture*, vol. 46, p. 1547-1554, 2006.

[BYR 93] BYRNE G., SCHOLTA E., "Environmentally clean machining processes – a strategic approach", *Annals of CIRP*, vol. 42, no. 1, p. 471-474, 1993.

[CAN 05] CANTERO J.L., TARDIO M.M., CANTELI J.A., MARCOS M., MIGUELEZ M.H., "Dry drilling of alloy Ti-6Al-4V", *International Journal of Machine Tools and Manufacture*, vol. 45, 2005, p. 1246-1255.

[DAV 06] DAVIM J.P., SREEJITH P.S., GOMES R., PEIXOTO C., "Experimental studies on drilling of aluminum (AA1050) under dry, minimum quantity of lubricant, and flood-lubricated conditions", *Proceedings of ImechE, Part B, Journal of Engineering Manufacture*, vol. 220, p. 1605-1611, 2006.

[DAV 07] DAVIM J.P., SREEJITH P.S., SILVA J., "Turning of brass using minimum quantity of lubricant (MQL) and flooded lubricant conditions", *Materials and Manufacturing Processes*, vol. 22, p. 45-50, 2007.

[DHA 06a] DHAR N.R., ISLAM M. W., ISLAM S., MITHU M. A.H., "The influence of minimum quantity of lubrication (MQL) on cutting temperature, chip and dimensional accuracy in turning AISI-1040 steel", *Journal of Materials Processing Technology*, vol. 171, p. 93-99, 2006.

[DHA 06b] DHAR N.R., KAMRUZZAMAN M., AHMED M., "Effect of minimum quantity lubrication (MQL) on tool wear and surface roughness in turning AISI-4340 steel", *Journal of Materials Processing Technology*, vol. 172, p. 299-304, 2006.

[DHA 07] DHAR N.R., AHMED M.T., ISLAM S., "An experimental investigation on effect of minimum quantity lubrication in machining AISI 1040 steel", *International Journal of Machine Tools and Manufacture*, vol. 47, p. 748-753, 2007.

[DIN 03] DINIZ A.E., FERREIRA J.R., FILHO F.T., "Influence of refrigeration/lubrication condition on SAE 52100 hardened steel turning at several cutting speeds", *International Journal of Machine Tools and Manufacture*, vol. 43, p. 317-326, 2003.

[GAI 08] GAITONDE V.N., KARNIK S.R., DAVIM J.P., "Selection of optimal MQL and cutting conditions for enhancing machinability in turning of brass", *Journal of Materials Processing Technology*, vol. 204, p. 459-464, 2008.

[GAI 10] GAITONDE V.N., KARNIK S.R., DAVIM J.P., "Study on the influence of MQL and cutting conditions on machinability of brass using artificial neural network", *International Journal of Materials and Product Technology*, vol. 37, no. 1-2, 2010, p. 155-171.

[HAN 03] HANYU H., KAMIYA S., MURAKAMI Y., SAKA M., "Dry and semi-dry machining using finely crystallized diamond coating cutting tools", *Surface and Coatings Technology*, vol. 173-174, p. 992-995, 2003.

[HEI 06] HEINEMANN R., HINDUJA S., BARROW G., PETUELLI G. "Effect of MQL on the tool life of small twist drills in deep-hole drilling", *International Journal of Machine Tools and Manufacture*, vol. 46, p. 1-6, 2006.

[IQB 08] IQBAL A., NING H., KHAN I., LIANG L., DAR N.U., "Modeling the effects of cutting parameters in MQL-employed finish hard-milling process using D-optimal method", *Journal of Materials Processing Technology*, vol. 199, p. 379-390, 2008.

[ITO 06] ITOIGAWA F., CHILDS T.H.C., NAKAMURA T., BELLUCO W., "Effects and mechanisms in minimal quantity lubrication machining of an aluminum alloy", *Wear*, vol. 260, no. 3, p. 339-344, 2006.

[KAM 07] KAMATA Y., OBIKAWA T., "High speed MQL finish-turning of Inconel 718 with different coated tools", *Journal of Materials Processing Technology*, vol. 192, p. 281-285, 2007.

[KAN 08] KANG M.C., KIM K.H., SHIN S.H., JANG S.H., PARK J.H., KIM C., "Effect of the minimum quantity of lubrication in high-speed end milling of AISI D2 cold-worked die steel (62 HRC) by coated carbide tools", *Surface & Coatings Technology*, vol. 202, p. 5621-5624, 2008.

[KEL 02] KELLY J.F., COTTERELL M.G., "Minimal lubrication machining of aluminum alloys", *Journal of Materials Processing Technology*, vol. 120, p. 327-334, 2002.

[KHA 09] KHAN N.M.A., MITHU M.A.H., DHAR N.R., "Effects of minimum quantity lubrication on turning AISI 9310 alloy steel using vegetable oil-based cutting fluid", *Journal of Materials Processing Technology*, vol. 209, p. 5573-5583, 2009.

[KIS 05] KISHAWY H.A., DUMITRESCU M., NG E.-G., ELBESTAWI, M.A., "Effect of coolant strategy on tool performance, chip morphology and surface quality during high-speed machining of A356 aluminum alloys", *International Journal of Machine Tools and Manufacture*, vol. 45, p. 219-227, 2005.

[KLO 97] KLOCKE F., EISENNBLATTER G., "Dry cutting", *Annals of CIRP*, vol. 46, no. 2, p. 519-526, 1997.

[KUM 00] KUMAR, P., BARUA P.B., GAINDHAR J.L., "Quality optimization (multi-characteristic) through Taguchi's technique and utility concept", *Quality and Reliability Engineering International*, vol. 16, p. 475-485, 2000.

[LIA 07] LIAO Y.S., LIN H.M., CHEN Y.C., "Feasibility study of the minimum quantity lubrication in high-speed end milling of NAK80 hardened steel by coated carbide tool", *International Journal of Machine Tools and Manufacture*, vol. 47, p. 1667-1676, 2007.

[LOP 06] LOPEZ DE LACALLE L.N., ANGULO C., LAMIKIZ A., SANCHEZ J.A. "Experimental and numerical investigation of the effect of spray cutting fluids in high speed milling", *Journal of Materials Processing Technology*, vol. 172, p. 11-15, 2006.

[LUG 97] LUGSCHEIDER E., KNOTEK O., BARIMANI C., LEYENDECKER T., LEMMER O., WENKE R., "Investigations on hard coated reamers in different lubricant free cutting operations", *Surface & Coatings Technology*, vol. 90, p. 172-177, 1997.

[MEN 06] MENDES O.C., AVILA R.F., ABRAO A.M., REIS P., DAVIM J.P. "The performance of cutting fluids when machining aluminum alloys", *Industrial Lubrication and Tribology*, vol. 58, no. 5, p. 260-268, 2006.

[MON 01] MONTGOMERY D.C., *Design and Analysis of Experiments*, John Wiley & Sons, New York, 1996.

[NAN 09] NANDI A.K., DAVIM J.P., "A study of drilling performances with minimum quantity of lubricant using fuzzy logic rules", *Mechatronics*, vol. 19, p. 218-232, 2009.

[OBI 06] OBIKAWA T., KAMATA Y., SHINOZUKA J., "High-speed grooving with applying MQL", *International Journal of Machine Tools and Manufacture*, vol. 46, p. 1854-1861, 2006.

[OBI 08] OBIKAWA T., KAMATA Y., ASANO Y., NAKAYAMA K., OTIENO A.W., "Micro-liter lubrication machining of Inconel 718", *International Journal of Machine Tools and Manufacture*, vol. 48, p. 1605-1612, 2008.

[OBI 09] OBIKAWA T., ASANO Y., KAMATA Y., "Computer fluid dynamics analysis for efficient spraying of oil mist in finish-turning of Inconel 718", *International Journal of Machine Tools and Manufacture*, vol. 49, p. 971-978, 2009.

[PHA 89] PHADKE M.S., *Quality Engineering using Robust Design*, Prentice Hall, Englewood Cliffs, NJ, 1989.

[RAH 01] RAHMAN M., KUMAR A.S., MANZOOR-UL-SALAM, "Evaluation of minimal quantities of lubricant in end milling", *International Journal of Advanced Manufacturing Technology*, vol. 18, p. 235-241, 2001.

[ROS 96] ROSS P.J., *Taguchi Techniques for Quality Engineering*, McGraw Hill, Singapore, 1996.

[SCH 97] SCHALKOFF R.B., *Artificial Neural Networks*, McGraw-Hill, Singapore, 1997.

[SIL 05] SILVA L.R., BIANCHI E.C., CATAI R.E., FUSSE R.Y., FRANCA T.V., AGULAR P.R., "Study on the behavior of the minimum quantity lubricant – MQL technique under different lubricating and cooling conditions when grinding 4340 steel", *Journal of the Brazilian Society of Mechanical Sciences and Engineering*, vol. 27, no. 2, p. 192-199, 2005.

[SIL 07] SILVA L.R., BIANCHI E.C., FUSSE R.Y., CATAI R.E., FRANCA T.V., AGULAR P.R., "Analysis of surface integrity for minimum quantity lubricant – MQL in grinding", *International Journal of Machine Tools and Manufacture*, vol. 47, p. 412-418, 2007.

[SOK 01] SOKOVIC M., MIJANOVIC K., "Ecological aspects of the cutting fluids and its influence on quantifiable parameters of the cutting processes", *Journal of Materials Processing Technology*, vol. 109, p. 181-189, 2001.

[SRE 00] SREEJITH P.S., NGOI B.K.A., "Dry machining – machining of the future", *Journal of Materials Processing Technology*, vol. 101, p. 289-293, 2000.

[SUN 06] SUN J., WONG Y.S., RAHMAN M., WANG Z. G., NEO K.S., TAN C.H., ONOZUKA H., "Effects of coolant supply methods and cutting conditions on tool life in end milling titanium alloy", *Machining Science and Technology*, vol. 10, no. 3, p. 355-370, 2006.

[TAS 08a] TASDELEN B., WIKBLOM B., EKERED S., "Studies on minimum quantity lubrication (MQL) and air cooling at drilling", *Journal of Materials Processing Technology*, vol. 200, p. 339 -346, 2008.

[TAS 08b] TASDELEN B., THORDENBERG H., OLOFSSON D., "An experimental investigation on contact length during minimum quantity lubrication (MQL) machining", *Journal of Materials Processing Technology*, vol. 203, p. 221-231, 2008.

[THE 09] THEPSONTHI T., HAMDI M., MITSUI K., "Investigation into minimal-cutting fluid application in high-speed milling of hardened steel using carbide mills", *International Journal of Machine Tools and Manufacture*, vol. 49, p. 156-162, 2009.

[WAG 98] WAGABAYASHI T., SATO H., INASAKI I., "Turning using extremely small amount of cutting fluid", *JSME International Journal, Series C*, vol. 41, no. 1, p. 143-148, 1998.

[WEI 04] WEINERT K., INASAKI, I., SUTHERL J.W., WAKABAYASHI T., "Dry machining and minimum quantity lubrication", *Annals of CIRP*, vol. 53, no. 2, p. 1-23, 2004.

[ZEI 06] ZEILMANN R.P., WEINGAERTNER W.L., "Analysis of temperature during drilling of Ti6Al4V with minimal quantity of lubricant", *Journal of Materials Processing Technology*, vol. 179, p. 124-127, 2006.

Chapter 4

Application of Minimum Quantity Lubrication in Grinding

This chapter deals with the application of minimum quantity lubrication (MQL) in grinding. This work aims to present some previous research results of the application of MQL in grinding by considering material to be ground (steels or ceramics) and type of grinding (surface, internal, and external cylindrical grinding).

4.1. Introduction

Grinding is the most common designation used to define the machining process that makes use of abrasive particles to promote material removal. It is traditionally considered as a finishing operation, capable of providing reduced surface roughness values along with narrow ranges of dimensional and geometrical tolerances [LEE 01, MAL 89].

The interactions between abrasive grains and the workpiece are highly intense, causing the required energy

---

Chapter written by Eduardo Carlos BIANCHI, Paulo Roberto de AGUIAR, Leonardo Roberto da SILVA and Rubens Chinali CANARIM.

per unit of volume of removed material to be almost completely transformed into heat, which is restricted to the cutting zone. The temperatures generated can be deleterious to the part, causing damage such as surface and sub-surface heating, allowing for surface tempering and re-tempering. It occurs in the formation of non-softened martensite, generating undesirable residual tensile stresses and reducing the ultimate fatigue strength of the machined component.

Moreover, uncontrolled thermal expansion and contraction during grinding contribute to dimensional and shape errors, leading mainly to errors in roundness. The grinding severity is limited by the maximum temperatures permissible during the process. When these are exceeded, they may lead to deterioration of the final quality [LIA 00, SIL 07]

To optimize the process, aiming for the control of thermal conditions, an increasing focus on proper tool selection emerges for each material to be ground. Also, the lubri-refrigeration method and types of cutting fluid applied have the role of reducing friction and heat, being responsible, as well, for expelling the removed material (chips) from the cutting zone. Adopting these procedures, it can be possible to machine with high material removal rates, obtain products with high dimensional and shape quality, and ensure the abrasive tool a greater life [WEB 95].

Cutting fluids in machining have the specific function of providing lubrication and cooling, thus minimizing the heat produced as a result of contact. Its drastic reduction or even complete elimination can undoubtedly lead to higher temperatures, causing reduced cutting tool life, loss of dimensional and shape precision, and even variations in the machine thermal behavior. An important and often forgotten function, which plays a decisive role in practice, is the ability to expel chips. When abrasive tools are used, a reduction in cutting fluid may render it difficult to keep the grinding wheel pores clean, favoring the tendency for clogging and thus contributing further to the aforementioned negative factors. However, it is noteworthy that the relative importance of each function also depends on the material being machined, the type of tool employed, the machining conditions, the surface finish, and the dimensional quality and shape required [TAW 08].

4.1.1. *Concern about cutting fluids*

In recent years, the unchecked consumption of non-renewable resources and immeasurable amounts of energy, air pollution, and industrial wastes have been the main focus of attention by public authorities. The environment has become one of the most important subjects within the context of modern life because its degradation directly impacts humanity. Driven by constant pressure from environmental agencies, politicians have drawn up strict legislation aiming to protect and preserve natural resources. These combined factors have led industrial sectors, research centers, and universities to study alternative production processes, which have resulted in the creation of technologies to minimize or avoid the production of environmentally deleterious residues and byproducts [SIL 07].

In metal cutting processes, the use of cutting fluids is the most common strategy to improve tool life, surface finish, and shape accuracy. It also makes breaking and transport of chips easier. However, there is often production of airborne mist, smoke, and other particulates in the workplace air. Those substances can cause several illnesses after prolonged exposures, such as dermatitis, malfunctions in the respiratory and digestive system, and sometimes even cancer [TAW 07, TAW 07a].

Emulsion-based cooling fluids for machining are still widely used in large amounts in industrial processes, generating high costs regarding consumption and disposal, not to mention the environmental harm. The growing need for an environmentally friendly production and accentuated rise concerning the costs involved in the fluid maintenance justify the demand for an alternative to those processes that apply it. In the last decade, however, the goal of research has been restricted to the minimum use of cooling and lubricating fluids in manufacturing processes. Dry machining and MQL have caught the attention of researchers and technicians as a promising alternative to the use of conventional fluids.

Confirming the trend for environmental concerns triggered by the unadvised use of cutting fluids in machining processes, as reported by several researchers and manufacturers, strong emphasis today is on technologies aimed at preserving the environment and conforming to the ISO 14000 standard. On the other hand, despite persistent attempts to completely eliminate cutting fluids, in many cases, cooling is still essential to achieve feasible tool service life and required surface qualities of the machined parts. This is particularly true when narrow tolerances regarding dimensions and shapes are required, or when materials of poor machinability are involved. MQL, in these cases, arises as an interesting option due to the combination of the functionality of cooling with an extremely low consumption of lubricant. These minimum amounts of oil suffice, in many cases, to reduce the tool friction while preventing the adherence of material [SIL 07].

4.2. Minimum quantity lubrication

Some limitations of dry machining can be overcome, in many cases, through the introduction of MQL systems, whose action is based on the application of 10 to 100 mL/h of lubricant on a compressed air jet, under pressures usually ranging from 4.0 to 6.5 kgf/cm^2 [SIL 07]. In this technique, the function of lubrication is ensured by the oil, whereas that of cooling is mainly ensured by the air. Although these advantages allow the foresight of a growing range of MQL applications, the influencing variables to be considered and its effects on the results have been the subject of very few studies [KLO 97, KLO 00, SIL 07].

MQL systems mainly use non–water-soluble cutting fluids, especially mineral oils. It should be considered that because of the reduced amounts of coolants used, the costs should not impede the use of high-technology compositions in the field of basic and additive oils. It is not recommended to use fluids that are designed for conventional systems, by virtue of the occurrence of atomization and vaporization, which are injurious to human health. Higher cutting speeds (which along with temperature cause problems of this kind) make the use of basic oils, with higher viscosity, and adaptations of additives (anti-mist) indispensable. The used lubricants should be environment friendly (free of solvents and fluorinated materials) and capable of high heat removal [HEI 98].

Many advantages follow from a comparison with conventional cooling [HEI 98, KLO 00, KLO 97]:

– The quotient concerning the amount of fluid used and the machined part volume is many times lower than that obtained by conventional cooling.

– Low consumption of fluid and elimination of a fluid circulation system.

– Filtering materials and devices along with maintenance recycling can be avoided.

– Low amount of oil remaining along with the machined chips does not justify its recovery.

– Machined parts are removed until almost dry, so in many cases it an unnecessary washing ensues.

– The application of biocides and preservatives can be eliminated because only the quantity to be used in a work shift should be added to the MQL system reservoir.

Regarding economical aspects, in comparison with conventional cooling, MQL causes additional costs concerning air pressurization and technological supports, which are intrinsically required to overcome its restrictions. For example, special techniques or devices for chip removal could be necessary and maybe the productivity would be reduced because of the thermal impacts on the machined components.

The oil vapor, mist, and smoke generated during MQL can be considered undesired by-products because these increase the air pollution of the workplace and thus becoming a factor of concern (being necessary, perhaps, an exhaustion system near the contact area). In pulverization, a compressed air line is used that functions intermittently during the process. These lines generate a level of noise that usually surpasses the limits allowable by legislation. Therefore, beyond affecting human health, the noise also pollutes the environment and prejudices the communication [KLO 00].

4.2.1. *Classification and design of MQL systems*

There are mainly three different types of MQL systems. First is the low-pressure pulverization, in which the coolant is suctioned by an air current and taken to the active surface as moisture. These systems are characterized by a flow rate of approximately 0.5 to 10 L/h. These are used mainly with emulsions, producing a notable optimization despite being able to dose only grossly [KLO 00].

The second type uses doser pumps with pulsatile feeding of a defined amount of lubricant, without air. The flow rates are adjusted in a range of 0.1 to 1 mL per cycle, for as much as 260 cycles per min. These systems are mainly applied in intermittent processes.

The third and widely used type is the pressurized system, in which the coolant is pumped to the nozzle through a distinct supply pipe. There occurs the mixing with compressed air, also supplied separately, so that the amounts of air and lubricant can be adjusted independently. This system is a particularly interesting alternative because it combines the functionality of cooling with extremely low consumption, ranging from 10 to 100 mL/h. At the same time, the coaxial moisture greatly prevents nebulosity.

MQL systems do not require much space to be installed on machine tools, being very versatile in its placement. These are easily integrated with the machines, making possible the addition of valves for a better control. These advantages make MQL a flexible system for grinding applications, along with other manufacturing processes [KLO 00].

The jet stability, which means the convergence of the moisture consisting of air and fluid, is very important for the practical application in machines for manufacturing, because it determines the distance from the nozzle to the active surface and therefore the danger of collisions between refrigeration systems and tools, machines, or the workpiece.

Oil mist formation is mainly caused by the function of MQL systems, which use the air-liquid moisture for lubri-cooling. All the oils used have low vapor pressure. In this way, the particles of oil mist are partially directed to the tool-workpiece interface, not allowing great dispersions. The damage to human health can be caused only by the particles, which are capable of entering and remaining in the lungs (known as "respirable particles"), with a diameter ranging from 0.5 to 5 µm. The larger ones do not pass through the nose, which acts like a filter, and the proportion of those that have a diameter larger than 0.5 µm is expelled in great quantity, along with the air. Only parts of the oil mist and smoke belong to the group of respirable particles, being considered relatively unhazardous, due to the lack of record of damages to the respiratory system, even with higher concentrations [HEI 98].

The function of an MQL system, however, is based exactly on generating the mist. In relation to its deleterious effects, there is a lack of proven facts. However, as the potential risk to human health is worrying, deep research in this area is necessary [HEI 98]. It can be emphasized that the nozzle design, the MQL equipment, and the lubricant used must allow a satisfactory formation of mist without dispersion. The lubricant used in the tests must also present excellent lubricity, contributing for the diminution of friction on the interface, thus resulting positively in the assessment of several parameters.

4.2.2. *MQL application in grinding*

A relatively little research has been conducted regarding the application of MQL in grinding. Some researchers investigated the effects of grinding parameters on AISI 4340 steel grinding by conventional lubrication and MQL. They found that the surface roughness, diametric wear, grinding forces, and residual stress improved when using the latter, due to optimum lubrication of the grinding zone, providing rather grain slipping at the contact zone [SIL 05, SIL 07]. Brunner showed that MQL grinding with ester oil at 4 mL/min (compared with mineral oil at 11 mL/min), when machining 16MnCr5 (SAE 5115) steel with microcrystalline aluminum oxide, reduced the process normal and tangential forces to one third, however, increasing the surface roughness by 50% [TAW 09].

Investigations by Brinksmeier confirmed these results and showed in addition that the type of coolant used during MQL grinding (ester oil or emulsion) can considerably influence the process result [TAW 09]. Hafenbraedl and Malkin [HAF 01] found that MQL provides efficient lubrication and reduces the grinding power and the specific energy to a level of performance comparable or higher than that obtained with conventional soluble oil (at a 5% concentration and a 5.3 L/min flow) while significantly reducing the grinding wheel wear. However, it presented slightly higher surface roughness values (R_a) [HAF 01, SIL 05, SIL 07].

The performance was also assessed when applying dry grinding. The results with the MQL technique were obtained in the internal cylindrical plunge grinding of AISI 51200 steel (quenched and tempered, detainer of an average hardness of 60 HRc), using conventional alumina wheels.

For the MQL technique, a precision doser providing ester oil at a specific flow rate (12 mL/h) was attached to the grinder. A nozzle mixed the oil with compressed air at a pressure level of 69 kPa, aiming to generate a thin mist. The application of ester oil for internal grinding was unsatisfactory, by virtue of the restricted area to the nozzle access. Its design was optimized to stay as close as possible to the inlet of grinding zone. Also, a cold air gun was used in an attempt to cool the workpiece. The cold air (−2°C) left the nozzle at a flow rate of 3 L/s and a pressure level of 7.6 bar. It was evaluated, subsequently, that the available amount of cold air would not be capable of providing significant cooling. However, the main disadvantage of MQL was the poor cooling, resulting in high temperatures and the thermal dilation of the workpiece.

Still in this line of thought, Baheti and others carried out some experiments with ester oil (10 mL/h) and cold air (−10°C at outlet) on surface grinding of carbon steel worpieces equipped a conventional wheel. The authors proved that the MQL technique presented lesser values of partition energy, temperature, and specific energy when compared with conventional cooling. When a comparison with soluble oil was made, MQL with cold air reduced the specific energy by 10% to 15%, the workpiece temperature by 20% to 25%, and the partition energy to the piece (fraction of grinding energy that is received as heat) by 15% to 20%.

Tests were carried out in different lubri-cooling conditions: liquid nitrogen; soluble oil (5% concentration); dry; ester oil; cold air (–10°C) at a flow rate of 990 L/min and pressure of 690 kPa; and cold air along with ester oil. Dry condition presented higher partition energy values, which was already expected. On the other hand, the application of liquid nitrogen provided higher specific energy values.

The researchers concluded that it is possible to eliminate or reduce the use of cutting fluid, contributing to clean manufacturing. Environmentally safe ester oil was capable of providing good lubrication and, when applied along with cold air, the cooling was more effective than that with soluble oil. Ester oil is classified as an unhazardous and non-carcinogenic substance. At the same time, MQL proves that it is a promising alternative to cutting fluids in grinding. Despite the fact that liquid nitrogen provided better cooling, it presents weak lubricity, which results in high values of specific energy [BAH 98].

Klocke and collaborators [KLO 00] presented the behavior of normal and tangential specific forces in external cylindrical plunge grinding when comparing the cooling by a shoe nozzle (24 L/min) and the MQL technique (215 mL/h), the latter resulted in a reduction in these forces. As far as microstructures are concerned, they revealed that no modifications happened, whatever the conditions employed. On the other hand, the MQL application presented the worst surface roughness values (R_z) when compared with wet grinding. The research proved, using results with defined geometry tools, that MQL can be used favorably in grinding processes [KLO 00].

However, extensive studies are necessary before this technology is applied industrially, mainly concerning the lubricant used. In this context, these are needed to verify the benefits and damage caused by this process, enabling it to become viable at the industrial scale. This research includes optimization of lubrication composition along with modifications in the design of grinders, abrasive tools, and monitoring strategies in order to adapt to different machining conditions.

According to Tawakoli [TAW 03], to make the MQL system favorable to grinding, certain developments are necessary on the following parameters: satisfactory systems for chip removal; optimized systems to supply cutting fluids in low flow rates; adjustment of the machining parameters, based on the complete understanding of MQL technology, for the chip thickness to reach an optimum value; reduction of friction; and optimized use of tools.

The results obtained by several researchers, until the present result, using the MQL technology with defined geometry tools, show the possibility of its application in many cases, contributing to clean manufacturing without harm to the environment or human health. They proved that MQL systems result in increased tool life, higher cutting speeds, better surface-finishing quality, and lesser damages to the workpiece. The MQL technology is perfectly qualified for manufacturing processes; however, it is indispensable to have an aggregation of effort between users, tools, and machine tool fabricants in order to obtain better results. It should be remembered, however, that despite all optimistic results with defined geometry tools, in relation to grinding, MQL is still far from its decisive implementation [TAW 03].

4.3. Results

The present section aims to present some previous research results of the application of the MQL technique in grinding. The authors are responsible only for the present results, recently obtained at the Laboratory of Abrasive Machining (LUA), located at UNESP – Univ. Estadual Paulista, Bauru campus, Brazil.

4.3.1. *Plunge external cylindrical grinding*

4.3.1.1. *Grinding of AISI 4340 steel with conventional wheels*

Some researchers have studied the grinding of tempered and annealed AISI 4340 steel. Classified as tempering steel, it is employed in the manufacture of pieces that require a good combination of mechanical strength and toughness [SIL 07].

The tests were carried out with aluminum oxide (Al_2O_3) wheels (355.6 × 25.4 × 127 mm; specification FE 38A60KV). The dressing conditions were kept constant, using a multi-granular *fliese* dresser that did not influence the output variables.

A series of preliminary tests was carried out to determine the best lubricant and compressed airflow, as well as the best choice of the various types of lubricants concerning the MQL application. From the seven types of lubricants subjected to preliminary testing, LB 1000 lubricant (supplied by the MQL equipment manufacturer) gave the best performance; therefore, all results reported here are related to this type of lubricant.

The equipment used to control the MQL was an Accu-Lube microlubrication system (ITW Chemical Products Ltd., Embu, São Paulo, Brazil), which uses an oil supply pulsating system, allowing the air and lubricant flows to be adjusted separately. The nozzle design allowed the compressed air to flow at a velocity close to the peripheral surface of the wheel (30 m/s). This is required to enable the mixture (lubricant plus compressed air) to penetrate the contact region, favoring lubrication and cooling. The lubricant flow was 40 mL/h. A flow meter and a pressure regulator equipped with a filter were installed to take precision measurements. The MQL system consisted of a compressor, a pressure regulator, a rotameter, a doser, and a spray nozzle. Figure 4.1 shows the nozzle designed and used in the testing of the MQL technique. It was placed at a distance of about 35 mm from the grinding wheel-workpiece interface.

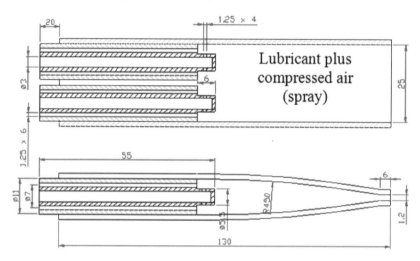

Figure 4.1. *Design of the MQL nozzle (dimensions in mm) [SIL 07]*

The main input parameters (wheel cutting velocity (V_s), plunge velocity (V_f), workpiece peripheral velocity (V_w), depth of cut (a), and spark-out time (t_s)) were selected on the basis of preliminary testing. The cutting conditions selected were: V_s = 30 m/s; V_f = 1 mm/min; V_w = 20 m/min (average velocity); a = 0.1 mm, and t_s = 10 s. These parameters were kept constant throughout the experiments.

A synthetic emulsion at a 5% concentration was used in the conventional cooling condition. The maximum flow supplied by the pump and the original nozzle was of 8.4 L/min.

The surface roughness was measured by adjusting a profilometer to a cut-off length of 0.8 mm. At the end of each test, the average values, R_a, were measured at three different points (approximately 120° equidistant from each other). Scanning electron microscopy (SEM) was used to analyze possible damage caused by thermal and mechanical stresses. The nominal values of residual stress were determined on the basis of the method of multiple exposition (sin 2Ψ), following the SAE J784a code.

4.3.1.1.1. Surface roughness

It is well known that surface finish can significantly affect the mechanical strength of components when they are subjected to dynamic loading. Figure 4.2 provides a comparison of the mean values of R_a (μm) with the Al_2O_3 wheel under conventional cooling with those obtained with MQL. The values were obtained after three 30-cycle plunge grinding stages. Six R_a measurements were taken in three different positions approximately 120° equidistant from each other.

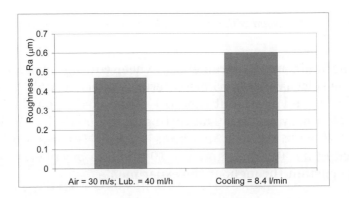

Figure 4.2. *Surface roughness after 90 cycles [SIL 07]*

An analysis of the results obtained indicates that the application of cutting fluid by the MQL technique gave higher performance to that of the conventional system because of the more efficient penetration of the fluid into the cutting region. The MQL technique led to lower surface roughness values, probably because of the more effective lubrication and cooling of the abrasive grains at the contact interface. This allows the chips to be expelled easily from the tool surface, resulting in a better surface finish. It should also be noted that no significant clogging of the grinding wheel pores was found with the MQL technique.

4.3.1.1.2. Residual stress

On the basis of a pre-analysis of the dry condition, it was also decided to measure the residual stress in order to compare the three conditions (conventional, MQL, and dry).

Figure 4.3 shows the values of residual stress for the samples ground with MQL, conventional cooling, and dry cooling. To identify how the grinding process affected the residual stress, it was also measured after turning, followed by heat treatment.

As indicated, grinding can lead to microstructural transformations because of high temperatures and displacement of the austenite in relation to the carbon, which helps diffusion. This may cause tensile or compressive stresses, depending on the material being ground and on the machining conditions.

According to Malkin [MAL 89], residual stresses can be caused by three factors: thermal dilation, microstructural transformations, and mechanical influence. Thermal dilations are proportional to the temperatures generated in the process, which are very high at the surface, gradually decreasing in the direction of the core. In grinding, when the heat source is active, the external layers dilate more than the internal ones, leading to residual compressive stresses on the surface. When it is no longer active (cooling), the external layer should contract more, which is not permitted by the lower layers. The mechanical influence, on the other hand, derives from the penetration of the abrasive grain into the piece [MAL 89].

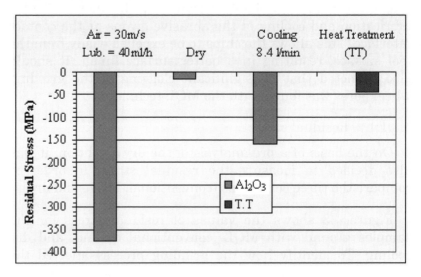

Figure 4.3. *Residual stresses at approximately 10 μm below the surface [SIL 07]*

Figure 4.3 shows that residual stresses are produced under both MQL and conventional conditions. Residual compressive stresses are considered beneficial for the mechanical properties, increasing the fatigue strength and the service life of components.

The MQL technique produced higher residual compressive stresses than those by the conventional technique, which is a positive aspect. Compared with dry grinding, there was an increase in the residual stress values obtained from tests performed under different conditions.

Compared with turning followed by heat treating, grinding with the MQL and conventional methods led to a significant increase in the residual compressive stress values, conferring the aforementioned characteristics on the mechanical properties of the machined part.

4.3.1.1.3. Microstructure analysis

Figures 4.4 and 4.5 are micrographs of sample cross-sections, illustrating the sub-surface alterations that took place in each condition. It can be noted that the sub-surface alterations were minimum for the various lubri-cooling methods, without significant differences between them.

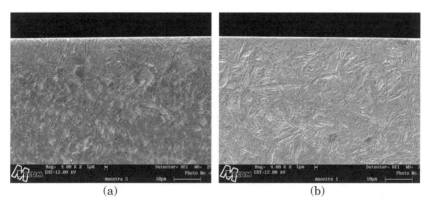

Figure 4.4. *Sub-surface microstructures obtained after 90 cycles [SIL 07]: (a) conventional cooling; (b) MQL (air = 30 m/s and lubricant = 40 mL/h)*

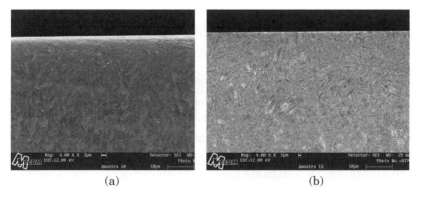

Figure 4.5. *Sub-surface microstructures obtained after 90 cycles [SIL 07]: (a) without cooling; (b) heat treatment*

4.3.1.1.4. Microhardness

Figure 4.6 represents the variation in microhardness as a function of the depth below the machined surface. To identify the influence of the grinding process on this variable, measurements of a sample taken directly from the heat treatment were made (i.e. without grinding), as shown in the following text.

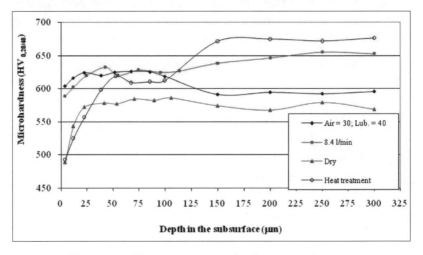

Figure 4.6. *Variation in microhardness as a function of subsurface depth [SIL 07]*

The results obtained for the two lubricating and cooling conditions did not indicate significant sub-surface modifications, confirming microstructural analyses. Steels usually undergo grinding after heat treatment. Depending on the temperature, grinding may cause annealing of the piece, thus softening the material close to the finished surface. Loss of surface hardness is a complex phenomenon related to annealing of the martensitic structure and to carbon diffusion and is dependent on the temperature and time involved in machining.

Malkin [MAL 89] stated that, in practice, it is interesting to combine this annealing behavior with a thermal analysis in order to predict hardness reduction. Experimental results demonstrate that high temperatures and long periods of exposure to such temperatures, at lower velocities or with longer contact lengths of the piece, led to greater losses in this property.

4.3.1.2. *Grinding of AISI 4340 steel with CBN wheels*

Three distinct lubri-cooling methods were used: the conventional method, using two circular nozzles; the MQL method, using a nozzle designed for this type of application; and the optimized method, based on the works of Webster [WEB 95a].

A SulMecânica RUAP515H CNC grinder was used for the process. Fifty workpieces were ground, 15 using MQL, 25 using the optimized method, 5 using the conventional method, and 5 under dry condition.

The test specimens were composed of quenched and annealed AISI 4340 steel, which is widely used in the manufacture of parts that require a combination of mechanical strength and toughness.

Vegetable oil emulsion (DMS 3200 F-1 manufactured by Shell) was used in the conventional and optimized methods, with 5% concentration and pH = 9. The cutting fluid used for the MQL method was Accu-Lube LB 1000 (ITW Chemical Products Ltd.), controlled microbiologically by ADEP 30 triazine bactericide.

The tests were carried out with a cubic boron nitride (CBN) wheel manufactured with vitrified bond and a 15% volume concentration of CBN, having the following characteristics: 350-mm external diameter, 127-mm internal diameter, 20-mm width, with 5-mm thickness, specification SNB151Q12VR2, having an open structure of fine hard grains.

The conventional nozzle used consisted of two misting nipples, each having an outlet diameter of 6.3 mm. The MQL system, on the other hand, consisted basically of a compressor, a pressure regulator, an airflow meter, a doser, and a nozzle designed for MQL in grinding.

The optimized system is characterized by high cutting fluid flow rate and pressure, consisting basically of an optimized nozzle designed from an MQL nozzle. A rectangular tube made of galvanized AISI 1010 sheet steel, which was shaped by fitting two U-channels into each other, was inserted into and welded to the nozzle outlet. Figure 4.7 illustrates the design of the optimized nozzle fabricated and used in this study.

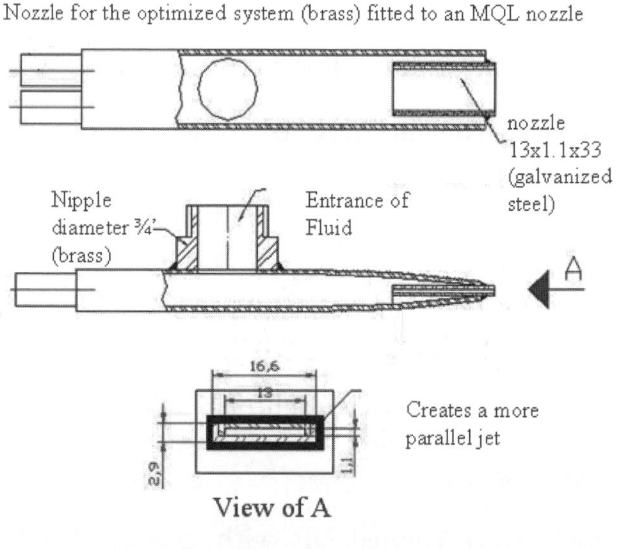

Figure 4.7. *Optimized nozzle design*

The machining conditions used in the tests were as follows: plunge velocity (V_f) = 1 mm/min; cutting speed (V_s) = 30 m/s; workpiece rotation = 204 rpm; penetration depth (a) = 0.1 mm; spark-out time (t_s) = 8 s; grinding width = 12 mm; grinding cycles = 100; maximum equivalent cutting thickness (h_{eq}) = 0.065 μm; minimum equivalent cutting thickness (h_{eq}) = 0.047 μm; and a fliese dresser.

Table 4.1 provides a list of cutting fluid velocities and flow rates in the optimized, MQL, and conventional conditions, respectively. It should be noted that the concentration of fluid used in the optimized and conventional techniques was kept constant at 5%.

Condition	Total cutting fluid flow (L/min)	Cutting fluid flow velocity V_i (m/s)
Optimized 30 m/s	26.3	30
Optimized 27 m/s	23.7	27
Optimized 25 m/s	21.9	25
Optimized 20 m/s	17.5	20
Optimized 15 m/s	13.2	15
	Total cutting fluid flow (L/min)	Compressed air exit velocity V_j, (m/s)
MQL 40 mL/h	0.00067	30
MQL 60 mL/h	0.00100	30
MQL 80 mL/h	0.00133	30
Conventional application	Total cutting fluid flow (L/min)	Cutting fluid flow velocity, V_j, (m/s)
Conventional	20	5.3

Table 4.1. *Speeds and flow rates used in the optimized, MQL, and conventional methods*

The tangential cutting force, F_{tc}, was measured from the electrical power consumed by the engine of the wheel axletree by monitoring the values of electrical voltage and current.

The acoustic emission signal was monitored in real time based on its mean quadratic value (root mean square, RMS), using a Sensis BM12 sensor. Parameters employed are as follows: signal gain: 3 dB; noise reduction: 30 dB; entrance gain: 3 dB; and time constant: 1 ms.

The mean surface roughness of the workpieces (R_a) was measured with a Taylor Hobson Surtronic 3+ rugosimeter, perpendicular to the grinding surface. The cut-off length was 0.8 mm, and a 2CR filter with phase correction was used. The diamond tip of the profilometer had a radius of 0.2 µm.

Scanning electron microscopy: a ZEISS DSM 960 scanning electron microscope was used with 2,000× magnification, applying the electron scattering technique, which offers a better view of the structure profile.

4.3.1.2.1. Tangential cutting force

Figure 4.8 shows the maximum mean tangential cutting forces of the five repetitions carried out for each of the MQL, optimized, and conventional conditions, allowing for evaluation and comparison.

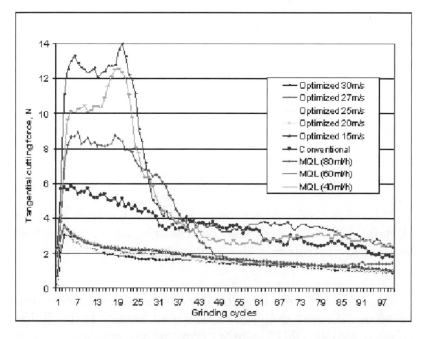

Figure 4.8. *Tangential cutting forces for each lubri-cooling condition*

An analysis of Figure 4.8 indicates that the cutting forces involved in MQL processes were much greater than those in the conventional and optimized systems.

The optimized system provided very efficient lubrication, reducing the grinding wheel wear and extending its service life. The high velocities involved in the optimized processes favored the entrance of lubricant at the workpiece-wheel interface, reducing the work of friction between the workpiece and the tool.

The conventional system showed an intermediary performance between the optimized and MQL systems in terms of cutting forces. In both the optimized and conventional systems, the apparent loss of sharpness was very subtle in the first 100 machining cycles (close to the 25th cycle in optimized lubrication). This finding indicates that this technique requires fewer wheel dressings when compared with the MQL method.

With regard to the variation in cutting fluid flow rates of 40, 60, and 80 mL/h for the MQL condition, the cutting force was found to decrease as the fluid flow increased. The condition that presented the best performance compared with conventional lubrication/cooling was the 80 mL/h of flow, which favored lubrication at the contact interface.

The optimized lubrication/cooling tests showed better results than those by the conventional and MQL methods, and the lowest tangential cutting forces observed with the optimized method were achieved with a cutting fluid flow velocity of 30 m/s.

The optimized technique presented lower tangential cutting forces than those by the conventional and MQL systems. However, the cutting force values obtained fell within an acceptable range for grinding processes.

4.3.1.2.2. Acoustic emission

Figure 4.9 illustrates the maximum mean acoustic emissions of the five repetitions carried out for each of the MQL, optimized, and conventional methods.

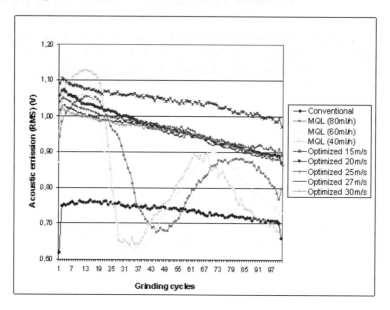

Figure 4.9. *Analysis of the effect of cutting fluid dispensing methods on the acoustic emission*

From a comparison of the three distinct lubri-cooling conditions, it can be seen that the conventional method behaved in a more consistent manner than that by the MQL method, generating the lowest values of acoustic emission.

The MQL test also showed that after the 20th grinding cycle, the minimum acoustic emission levels were close to the blank emission levels. No significant differences were detected in the minimum emissions as a function of the cutting fluid flow applied in each MQL condition.

The acoustic emissions in the optimized and conventional methods were more constant and varied little, indicating that these forms of lubrication could maintain the sharpness of the grinding wheel throughout the grinding cycles.

According to Inasaki [INA 91], there is a relationship between specific grinding energy and acoustic emission values, which was confirmed by their tendency to minimize as the cutting fluid flow rates increased.

4.3.1.2.3. Surface roughness

Figure 4.10 shows the mean surface roughnesses for each condition tested.

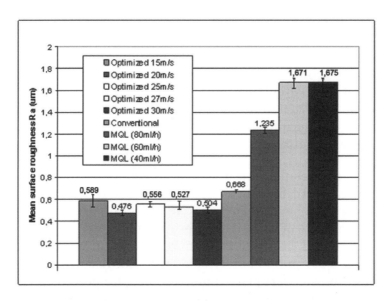

Figure 4.10. *Mean surface roughness for each lubri-cooling condition*

An analysis of the results indicates that the surface roughness values were generally lower in optimized lubrication/cooling, followed by the conventional method, and finally by MQL. The lowest values for MQL were obtained by a lubricant flow rate of 80 mL/h, thus confirming that the larger amount of fluid led to lower surface roughness values because of the greater lubrication it provided.

The conventional lubrication/cooling method resulted in lower surface roughness values than those by the MQL technique, but the values obtained by a lubricant flow of 80 mL/h are lower than those of the majority of industrial applications.

The best optimized situation (fluid flow velocity of 20 m/s) with respect to the mean surface roughness, R_a, of the workpiece was 71.7% lower than that of the best MQL condition (Q = 80 mL/h) and 47% lower than that of the conventional cutting fluid dispensing method.

The lowest values attained with the optimized method were obtained at higher cutting fluid flow rates, thus confirming that a larger quantity of fluid allowed for lower surface roughness values because of the greater lubrication provided. Higher cutting fluid flow rates allow for more rapid chip removal, contributing to a better finish. The differences between the optimized conditions were minor but showed a tendency for better quality at higher cutting fluid flow rates.

From the results of tangential cutting forces and acoustic emission, it can be concluded that the MQL system with a compressed air exit velocity of 30 m/s, CBN grinding wheel containing 15% of abrasive material, and a cutting velocity of 30 m/s should be used for a maximum equivalent cutting depth (h_{eq}) of approximately 0.060 μm. Because this variable depends on the workpiece diameter, the plunge velocity, and the cutting velocity, these parameters should be suitably adjusted to enable the grinding wheel to cut (lowest removal per abrasive grain) and the MQL system to remove the chips from the wheel pores.

The conditions of cutting fluid and compressed airflow rates used with the MQL technique did not cause misting, giving the operator a good view of the process and favoring environmentally correct machining.

4.3.1.2.4. Microstructural analysis

The surface integrity of a workpiece is an extremely important factor, and damage of the surface of a material may affect it significantly, causing degradation of the properties of wear, corrosion resistance, crack nucleation, and propagation and acceleration in the fatigue process of components. The surface integrity of the workpiece is affected principally by the temperature produced by the grinding process, which may lead to its thermal damage.

The conventional method provides efficient lubrication and cooling of the workpiece, without allowing it to sustain damage. The same holds true for optimized lubrication/cooling. In addition, the contact time of abrasive grains and the cooling time are very short, thus not leading to significant differences at the sub-surface.

The MQL condition of 80 and 60 mL/h did not lead to substantial sub-surface alterations of the microstructure when using the MQL technique. The microstructure depicted in Figures 4.11 to 4.14 corresponds to the MQL condition of 40 mL/h. In this condition, it was possible to detect significant sub-surface changes, such as cracks and surface burn, in the microstructure.

An analysis of the microstructure indicated that the various conditions tested using optimized and conventional methods presented satisfactory results. In other words, no significant microstructural changes or surface damage of the workpiece were found after grinding, thus improving the component properties of corrosion resistance and abrasive strength, as well as enhancing its fatigue strength. The only exception was the situation in which a cutting fluid flow rate of 40 mL/h was used by the MQL technique, which generated cracking and surface hardening, albeit without visually detectable burn, mainly because of the smaller amount of lubricant delivered, causing generation of heat in large quantities.

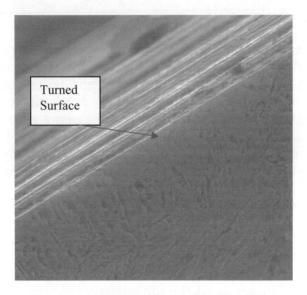

Figure 4.11. *Micrograph of a turned specimen (2,000× magnification)*

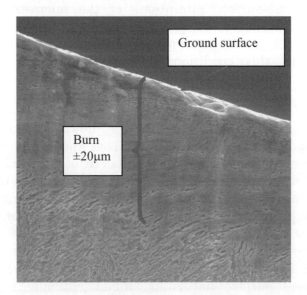

Figure 4.12. *Micrograph of a test specimen with dry grinding (2,000× magnification)*

Figure 4.13. *Micrograph of a test specimen subjected to MQL under cutting fluid flow of 80 mL/h (2,000× magnification)*

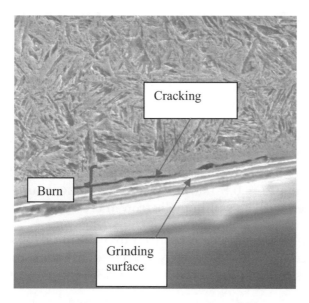

Figure 4.14. *Micrograph of a test specimen subjected to MQL under cutting fluid flow of 40 mL/h (2,000× magnification)*

4.3.1.3. *Grinding of advanced ceramics with diamond wheels*

The experiments were performed in a SulMecânica RUAP515H CNC cylindrical grinder, equipped with a computerized numerical control (CNC) system.

Cylindrical workpieces of commercial alumina (96% aluminum oxide and 4% bond oxides as SiO_2, CaO, and MgO) were ground. The apparent density of this material was 3.7 g/cm^3.

The process used a resinous bond diamond grinding wheel, with 350 mm (outside diameter) × 15 mm (width) × 5 mm (layer), internal diameter = 127 mm, specification D107N115C50 (Nikkon Cutting Tools Ltd., Suzano-SP, Brazil).

The cutting fluid used was a semi-synthetic ROCOL 4847 Ultracut 370 emulsion of 5% concentration, which already contained anti-corrosives, biocides, fungicides, and other additives. To control MQL, an Accu-Lube microlubrication system was used (ITW Chemical Products Ltd.), which uses an oil supply pulse system and allows the air and lubricant flow rates to be adjusted separately.

Measurements of noise emissions were made by a Sensis DM12 acoustic emission sensor, positioned at the head of the mobile near the tailstock. The roundness was measured on a Taylor Hobson Talyrond 31c equipment. The surface roughness was measured with a Surtronic 3+ profilometer (cut-off length = 0.8 mm). The microstructure analysis was performed with a scanning electron microscope.

The wheel wear was measured by printing its profile on a 1010 steel workpiece properly prepared and then the data of variables could be gathered with a TESA comparator gauge.

The tests were performed under the following machining conditions: plunge speed (V_f) = 1 mm/min; wheel peripheral speed = 30 m/s; depth of cut = 0.1 mm; spark-out time = 5 s; fluid flow rate in conventional cooling = 22 L/min; flow rate of the fluid in MQL = 100 mL/h and air pressure of 8 bar; outlet air velocity in a MQL nozzle = 30 m/s; and 13 workpieces per test.

The following three feed rates were chosen: 0.75, 1, and 1.25 mm/min corresponding to the respective equivalent thicknesses of cut h_{eq1} = 0.0707 mm, h_{eq2} = 0.094 mm, and h_{eq3} = 0.118 mm, respectively.

4.3.1.3.1. Acoustic emission

Figure 4.15 presents the results of acoustic emission (RMS), expressed in volts, according to the number of ground pieces.

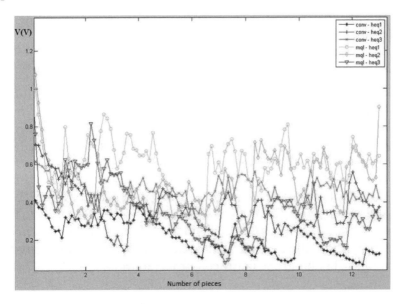

Figure 4.15. *Acoustic emission results*

The values in Figure 4.15 indicate no significant differences in relation to acoustic emissions. It can be seen that the condition that showed lower values was the conventional cooling with h_{eq1} (smaller equivalent thickness of cut), whereas the condition that showed higher values was the MQL technique with h_{eq1}.

One explanation for these phenomena is the small influence of the equivalent thickness of cut in MQL caused by other significant variables, such as the thermal dissipation of the cutting region. Since this method dissipates less heat, its removal mainly occurs by the thermal conduction of the grinding wheel, which is constant for all tests. As the equivalent thickness of cut is determined by the feed rate, the higher thickness of cut provides greater contact area, thus causing more heat conduction.

This type of conclusion can be drawn just because the workpiece has small thickness with respect to the thickness of the grinding wheel. For workpieces with greater thickness, the thermal conduction of the wheel can be limited.

4.3.1.3.2. G-ratio

Figure 4.15 presents the G-ratio for each equivalent thickness of cut and lubri-cooling condition. This value was calculated by measuring the wheel wear and volume of worn material. The first could be measured because of the greater width (15 mm) with respect to the workpiece (4 mm).

From the analysis of Figure 4.16, it can be noticed that the higher values for the G-ratio were obtained for conventional cooling. One possible reason is the lower heat dissipation in the cutting region caused by MQL, resulting in the loss of bond resistance, thus more wearing of the grinding wheel.

It can be also seen that for the conventional cooling, the equivalent thickness of cut is a great influence on wheel wear, therefore the G-ratio. The higher its value, the more accentuated the wear, consequently providing lower values for the G-ratio.

Application of MQL in Grinding   143

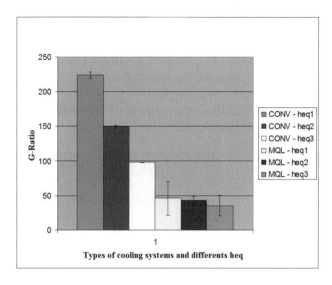

Figure 4.16. *G-ratio*

For the MQL technique, the equivalent thickness of the cut could not influence effectively in the G-ratio. This can be explained by other factors that probably prevailed in the wear, that is, the lower heat dissipation in the cutting zone, making the influence of equivalent thickness of cut almost imperceptible.

4.3.1.3.3. Scanning electron microscopy

Figure 4.17 presents the results of the SEM obtained for the conditions of lubri-cooling with 1,000 times magnification.

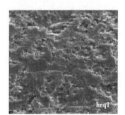  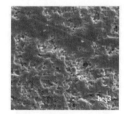

Figure 4.17. *SEM for conventional cooling for h_{eq1}, h_{eq2}, and h_{eq3}*

In the conventional cooling, the fragile mode of material removal takes place. The tendency to ductile mode removal increases, as does the equivalent thickness of cut, providing an improvement in the workpiece finishing.

Figure 4.18 presents the results of the MQL technique, with 1,000 times magnification.

  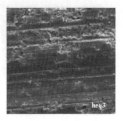

Figure 4.18. *SEM for cooling the MQL system for h_{eq1}, h_{eq2}, and h_{eq3}*

It can be noticed that the predominant mode of material removal by MQL is the ductile mode, which provides optimal conditions for surface finish with mechanical strength because of the reduction in microfractures, which are responsible for stress concentrators. From Figure 4.18, it can be seen that the lower the equivalent thickness of cut, the more ductile the process of material removal is.

The better surface characterization by MQL than with the emulsion used in the conventional cooling may be explained by the greater power of the lubricating oil used.

4.3.1.3.4. Roundness

Figure 4.19 shows an evolution of the roundness for all conditions tested.

It can be noted that only for the more severe condition of MQL lubrication, the roundness has increased dramatically.

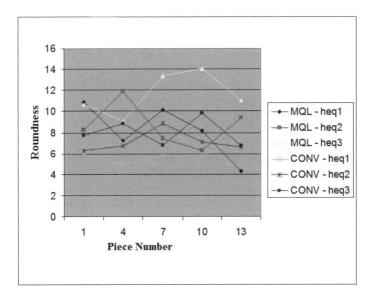

Figure 4.19. *Evolution of roundness errors*

Analyzing the results obtained as a whole, the values for less severe conditions by MQL did not differ significantly.

For the roundness for all conditions tested, there were no significant differences between both methods for h_{eq1} and h_{eq2}.

4.3.1.3.5. Surface roughness

Figure 4.20 shows the comparison results of the average surface roughness (R_a) for the conventional lubri-cooling and MQL (in micrometers). The values shown are averages of five measurements at different positions, for each of the three tests, with their respective standard deviations.

In general, the values were lower for the conventional lubri-cooling method than for the MQL method, possibly due to the better chip removal from the cutting zone by conventional cooling. When applying the MQL technique, a paste of fluid and chips was formed even with compressed air at high speeds. This considerably affected the values of surface roughness.

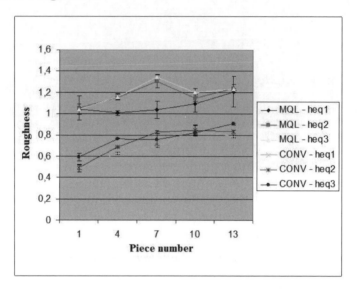

Figure 4.20. *Evolution of surface roughness during the tests*

The lower values for MQL are observed in the lowest values of h_{eq}, demonstrating that the smaller thickness of cut allows smaller values of surface roughness because of lower material removal rate and greater lubrication achieved.

The surface roughness is mainly influenced by the lubrication condition. The emulsion presents the characteristics of less lubrication but more cooling, thus affecting this variable.

4.3.2. *Internal plunge grinding*

4.3.2.1. *Internal plunge grinding of SAE 52100 steel*

The experimental set-up consisted of a SulMecânica RUAP 515 H CNC cylindrical grinder, equipped with accessories such as high rotation head for internal cylindrical grinding, where it was fixed on the wheels; a support to fix the workpiece, manufactured to eliminate the localized stresses on the clamp fixation; optimized and conventional nozzles; MQL pneumatic system; and a fliese dresser. The workpieces were made of quenched and softened hollow SAE 52100 steel cylinders (internal diameter = 38.0 mm), with an average hardness of 60 HRc.

The wheels used were of conventional white alumina (Al_2O_3), with a specification of 38A100MVHB; a conventional white alumina wheel with YT treatment (addition of sulfur), 38A100MVHBYT; a conventional white alumina wheel with "12" treatment (addition of natural waxes), 38A100MVHB12; and a ceramic (seed-gel alumina) 5ES100M10VHB wheel. All wheels had 12.0 mm of external diameter and were provided by Saint-Gobain Abrasives Ltd. According to the manufacturer, the main objective of the treatments is to increase contact lubrication, avoiding burns and improving the product final quality. Also, the wheel impregnated with sulfur is undesirable at the workplace; therefore, the wheel with 12 natural waxes is environmentally unharmful. Both have higher cost than that of the conventional wheel, and for this reason a study of their viability should be carried out.

In the present work, the workpieces were ground, being subjected to three different flow rates, for the optimized lubri-refrigeration method: 21, 16, and 12 L/min, which correspond to outlet velocities of 27, 20, and 15 m/s, respectively. The conventional method was used with a flow rate of 10 L/min; MQL, a pressure level of 8 kgf/cm^2 and a flow rate of 80 mL/h were kept constant. Before starting each test, all tools were dressed up.

For each test, 180 cycles of 8-μm radial feed were executed, removing a length of 1.44 mm from the workpiece internal radius. After the tests, these lengths of cut were cleaned and the output variables (e.g. surface roughness, roundness errors, and diametral wear) were measured.

4.3.2.1.1. Surface roughness

Figure 4.21 illustrates the surface roughness values (in µm) for each condition of lubri-refrigeration and wheel used.

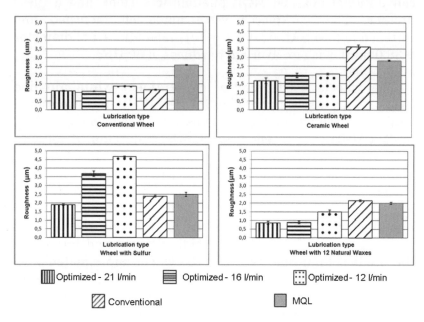

Figure 4.21. *Surface roughness versus lubrication type*

Analyzing, *a priori*, the results for the conventional wheel, it can be verified that the flow rates of 21 and 16 L/min provided a more efficient penetration of the fluid on the contact zone, due to the disruption of the aerodynamic barrier, by virtue of the greater closeness to the cutting speed (30 m/s), thus minimizing the friction, heat generation, and cutting forces.

When the flow rate of 12 L/min was used, the efficiency of refrigeration was substantially reduced, causing an increase in surface roughness; it must be taken into account that the fluid jet is restricted, limited to the outlet area of the optimized nozzle (with respect to the conventional wheel).

However, the result of this flow rate was more satisfactory than that when MQL was used, because the chip expulsion from the contact zone was not ensured effectively, thus presenting the worst results of all tests.

For the ceramic wheel, it can be verified that optimized method, independently of the flow rate, gave the best results, despite being worse than those obtained with the conventional wheel, because of the same aforementioned causes. Restricting the analysis to the optimized method, it can be noted that the higher flow rate (21 L/min) provided the lower values of surface roughness; there is no significant difference between those obtained from the lower ones when observing the standard deviations.

For all tools, the optimized lubri-refrigeration method when properly applied using outlet jet speed close to the cutting velocity provided the lower surface roughness values. Such fact can be explained by the higher efficiency of heat removal and reduction in friction and cutting forces. However, when flow rates of 12 and 16 L/min were used, the method did not always ensure better performance because of the fluid jet restriction, which prevents proper penetration on the contact area and ineffective aerodynamic barrier disruption, as occurred in the wheel impregnated with sulfur.

When the MQL technique was used, the results obtained were always worse in comparison with the higher flow rate optimized method. This is due to the inefficient removal of chips because they cluster with the lubricant and become a type of paste that remains on the cutting zone and scratches the workpiece surface, deteriorating the intended finishing. Independently of the wheel used, MQL provided similar results for surface roughness.

In comparison with the conventional wheel, the ceramic wheel presents a characteristic that is typical of high material removal rates because of its sharpened grains (according to the manufacturer); the obtained surface roughness for the same lubri-refrigeration method and machining parameters always had higher values than those with the conventional wheel.

The wheel containing sulfur did not provide good results in comparison with the conventional method, because of already impregnated pores, obstructing the proper exit of chips, diminishing the cutting capacity of the tool, and prejudicing the results.

The wheel with 12 natural waxes, despite also having impregnated pores, provided similar results in comparison with the conventional wheel (except when used with the conventional refrigeration); however, its use is not justified because of its high cost in comparison with the untreated alumina wheel.

4.3.2.1.2. Roundness errors

Figure 4.22 illustrates the roundness errors (in µm) for each lubri-refrigeration method and wheel used.

In a way similar to what occurred with surface roughness, it can be verified that for the conventional wheel, a reduction in roundness errors occurred when higher fluid outlet velocities were used. This reduction was significantly accentuated with a flow rate of 21 L/min. This is due to the reduction in friction between the wheel and the workpiece, minimizing cutting forces, thus resulting in a lesser bending on the axis that supports the wheel. In this way, a lower vibration is produced on the grinder system.

It can be found on analyzing the ceramic wheel that the best result obtained was with the higher flow rate (21 L/min) from the optimized method because of the efficiency in disrupting the aerodynamic barrier, better lubrication of the contact zone and reduced cutting forces, diminishing the vibration. When using 16l/min and 12l/min, it can be perceived a significant increase, in relation even to conventional and MQL methods. Such fact is caused by the difficulty in introducing the fluid on the contact zone, not only by virtue of the presence of an aerodynamic barrier, but the difficulty of the fluid in accessing the great contact area between workpiece and wheel, which contributes for a worse lubrication, and consequent loss of shape and increase in the system vibration.

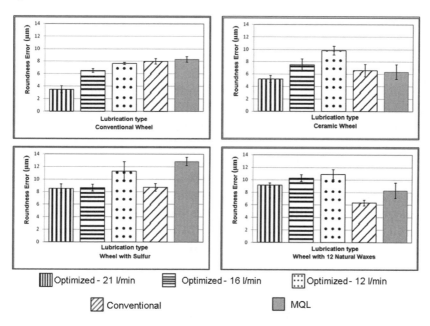

Figure 4.22. *Roundness error versus lubrication type*

For MQL, it can be noticed that the conventional wheel provided the worst results as a result of the higher wheel loss of shape, caused by friction, because in this lubri-refrigeration method the lubricant did not behave as in the other conditions. In the case of the ceramic wheel, the results obtained for MQL are similar to the conventional method because in this case (according to the manufacturer), the wheel has a better shape maintenance, thus reducing the roundness errors.

For the optimized method, the conventional wheel presented the best results when compared with the ceramic wheel, proving that the more efficient the lubrication, the better is the performance regarding roundness errors. However, with other methods (conventional and MQL), the ceramic wheel presented the best results, by virtue of its shape maintenance characteristic (as previously explained), which is non-existent in the conventional wheel.

In the case of the wheel containing sulfur, the optimized method with higher flow rates did not present difference in comparison with the conventional wheel, and all values obtained with this abrasive wheel were significantly higher than those obtained by other methods. This occurs because of sulfur-impregnated pores so that the chips cannot find a proper exit surface, thus increasing the cutting forces. With this increase, there is an elevation in vibration and roundness errors. In other words, the lubrication effect generated by sulfur could not remove the effect caused by obstruction of the wheel pores. This is evidenced clearly on the tests in which poor lubrication conditions were used, which presented the higher roundness error values.

The wheel containing 12 natural waxes, despite having its pores impregnated, presented good performance with the application of the conventional and MQL methods. This occurs because this wheel can combine a satisfactory lubrication effect, by virtue of the waxes, with a better exit surface for the chips, thus providing better results on the more critical conditions of lubri-refrigeration. For the optimized method, an increase in the roundness errors can be noted with the reduction in fluid jet outlet velocity, by virtue of the stress increase caused by worse penetration in the contact zone, and the obstruction of pores resulting in worst results.

4.3.2.1.3. Wheel diametral wear

Wheel wear occurs due to three main factors: agglutinant wear, abrasive wear, and grain friability (capacity of the grain of generating new cutting edges when subjected to stresses). This variable has a great significance in grinding since higher wear mean shorter wheel life cycles.

Figure 4.23 illustrates the wheel diametral wear for each lubri-refrigeration method and wheel used.

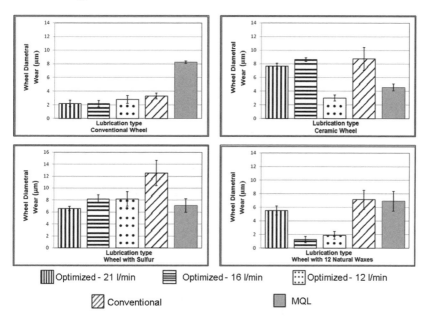

Figure 4.23. *Diametral wheel wear versus lubrication type*

It can be verified that for the conventional wheel, best results were obtained with flow rates of 21 and 16 L/min; there exists a tendency for worst results with less efficient lubrication methods because of an increase in cutting forces and consequently in wheel wear.

With the results from ceramic wheels and those treated with 12 natural waxes, it was not possible to notice any tendency concerning the machining parameters and the other relevant factors in grinding operation. This is mainly due to the way the tool wears off, as well as in the case of sulfur wheel, for which an irregular wear occurs due to vibration generated by high cutting stresses, shaft bending, and great contact area. This fact is presented clearly by the high standard deviation values found in some results.

It can be emphasized by the difficulty in measuring this variable, since it was performed by visual access in a computer-aided design software, and the lack of a more efficient method to obtain the wheel wear. In addition, the tool will always present an irregular wear on its surface, adversely affecting the accuracy in measurements when used the applied method.

4.3.3. *Surface grinding*

4.3.3.1. *Grinding of AISI 4340 steel with CBN wheels*

The equipment set consisted of a SulMecânica 1055E surface grinder, a vitrified CBN grinding wheel (350-mm external diameter, 127-mm internal diameter, 20-mm width, with 5-mm thickness), specification SNB151Q12VR2; AISI 4340 steel workpieces, tempered and quenched (54 HRc average hardness), 100 × 200 × 10 mm dimensions (rectangular shape). The dressing operation was kept constant, using a fliese dresser that did not influence the output variables.

The cutting fluid used in the conventional and optimized methods with a Webster nozzle (1999) was semi-synthetic soluble oil, 5% concentration (as recommended by the manufacturer). In the conventional method, this fluid is adequately applied on the contact interface, with a flow rate of approximately 27.5 L/min. In the optimized method, a pump makes it possible to apply the fluid at an outlet speed of 32 m/s, the same as the grinding wheel. In this case, the pressure used was about 8 bar and flow rate is the same as that in the conventional method (27.5 L/min).

The MQL system consisted of a compressor, a pressure regulator, an air flow meter, a dosing device, and a nozzle. This set allows a fine adjustment of lubricant/air volume separately with a needle-type register. The cutting fluid used was biodegradable vegetal oil with additives for extreme pressure, severe operations, and anti-oxidants. In this experiment, the airflow was maintained at pressure of nearly 8 bar and a cutting fluid flow rate of 100 mL/h.

The optimized method, using a Webster nozzle, requires a special geometry with rounded surfaces, which can reduce turbulence flow of the cutting fluid. This nozzle was built as that reported by the author [WEB 99].

Preliminary experiments were carried out to define grinding parameters. Maintaining the workpiece speed at 0.0033 m/s and the wheel peripheral speed at 32 m/s, both being constant, the depth of cut varied ranging from 0.02 to 0.05, and 0.08 mm.

The end criterion was the specific volume of the material removed, which in this case is 5,000 mm^3/mm. Since there were three different depths of cut, the time to obtain this volume was different in each situation.

The tangential cutting force was measured by electric power consumed by the engine of the grinder spindle, using an electronic module and a board of data gathering, manipulated by a data gathering program, based on the National Instruments® LabView software.

Surface roughness data were obtained by measuring R_a with a Taylor Hobson Surtronic 3+ rugosimeter at each 1,000 mm^3/mm of the removed material.

The wear measurement for the wheel was performed with a specimen of AISI 1045 impress steel, which was measured by a Tesa Micro-Hite 3D direct computer control system.

Vickers microhardness was also determined by a Buehler 1600-6300 tester.

For each depth of cut, three different lubri-cooling methods were tested, the conventional, MQL, and optimized applications. For each method, the experiments were repeated three times, with a total of 27 experiments being carried out.

4.3.3.1.1. Tangential cutting force

Figure 4.24 shows the tangential cutting force behavior for the three depths of cut experiments. It can be observed that in the MQL technique with 0.02 and 0.05 mm depths of cut, the cutting force is smaller than that by the conventional and optimized methods. This can be explained by the fact that MQL, under these conditions, provides a more effective lubrication in the cutting zone, overcoming the air barrier generated by the grinding wheel. This way, the proportion of forces spent on the friction is minimized. Similar condition is observed in the optimized method, which shows behavior quite similar to that of MQL.

At 0.08 mm depth of cut, the force observed in the MQL technique was the highest, the grinding condition being the most severe among the ones analyzed, which can be confirmed by an increase in the cutting force. The fact that MQL forces are greater refers to the difficulties with efficient lubrication. Under these severe conditions of cutting, high stresses were present. Besides the inefficient lubrication, the cutting zone cooling by the airflow could not minimize the excessive heat generation. In this case, the optimized method seems to be more efficient, allowing better penetration in the cutting region and the removal of the excessive heat generated.

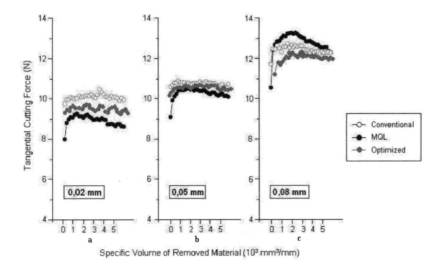

Figure 4.24. *Tangential cutting force: (a) 0.02 mm; (b) 0.05 mm; and (c) 0.08 mm*

The optimized method presented good results because of its capability of cooling the cutting zone, which relies on the way the cutting fluid is applied. The high speed enables the fluid to penetrate the cutting zone, disrupting the air barrier generated by the grinding wheel. This method behaves in a consistent manner with the increase in the depth of cut. The conventional method, on the other hand, presents higher values caused by the air barrier, which the abundant flow of fluid cannot surpass efficiently.

It is possible to conclude that with an increase in the depth of cut, the average values of tangential cutting force tend to increase as well. This is a consistent presumption, since it produces a bigger contact area between the grinding wheel and the workpiece, and although the number of grains in contact is also bigger, the strength to remove a bigger volume of material also increases, which consequently produces higher cutting forces.

Observing the cutting force behavior, which varies along the experiment, it can be concluded that it is due to the CBN grains friability. Along the process, the grains are worn out, losing their cutting edges, thus becoming flatter. This produces higher cutting forces because of a decrease in their cutting capacity and an increase in the drag process. These worn out grains are released from the matrix, renewing the cutting edges. CBN grinding wheels have the capacity of self-sharpening, which strengthens their quality characteristic.

4.3.3.1.2. Surface roughness

It is important to analyze the surface roughness since the surface finishing affects the fatigue strength of the workpieces significantly when they have less strength. The surface roughness is still linked to lubrication and depends mainly on the abrasive grain size present in the wheel, on the dressing conditions, and material removal rate [MAL 89].

Figure 4.25 presents the surface roughness values (in μm), resulting from the arithmetic mean of all experiments with the same depth of cut.

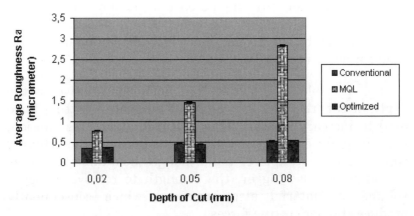

Figure 4.25. *Comparison of surface roughness among the methods tested*

Figure 4.25 shows that all surface roughness data gathered in the conventional and optimized lubri-cooling methods are lower than the values obtained by the MQL method. At 0.08 mm depth of cut, data in MQL are almost six times higher than those obtained by the other methods.

According to Malkin, the acceptable range for average surface roughness in a grinding process is 0.2 to 1.6 µm. Therefore, all values obtained with the conventional and optimized methods are within the tolerance range, being quite satisfactory. The values are very similar to these methods, so it is possible to conclude that both methods remove chips efficiently and guarantee the characteristic surface quality of the grinding process [MAL 89].

In the MQL method, however, surface roughness data obtained at 0.08 mm depth of cut are outside the required range, which disqualifies this condition concerning the surface quality. At the best condition (0.02 mm), the values produced by the MQL method were higher than those obtained by the other methods. Nevertheless, the MQL method is a considerable alternative, even for finishing. For 0.05 mm, the MQL method lies within the limit of the considered range, which means that the surface quality has not been provided. In this case and for 0.08 mm depth of cut, the MQL technique can be disqualified.

The reasons of these variations can be explained by the conventional and optimized methods that allow the removal of the chips in the cutting zone, improving the lubrication in the region and minimizing the friction between the tool and the workpiece. The MQL method, which is inefficient in these tasks, causes the chips to remain in the cutting zone; the bigger the chip, more difficult is to remove it and worse surface quality is generated, which can be corroborated by the high values of surface roughness. The airflow is efficient in removing the chips with 0.02 mm depth of cut. However, its efficiency significantly decreases with the increase in the depth of cut.

Although MQL requires a lower cutting force, the workpiece surface roughness is higher. The explanation of this fact comes from the grinding wheel topography. The CBN wheel lodges the thin chips produced in the MQL method, resulting in lower forces and medium surface roughness. However, the increase in the depth of cut increases the chip size, which makes it more difficult for the chips to settle in the wheel porosity, thus resulting in depleted surface quality.

4.3.3.1.3. Grinding wheel wear and G-ratio

Grinding wheel wear is an extremely important variable in grinding process, since it is related to the wheel life. Figure 4.26 presents the diametric wear of the grinding wheel at each depth of cut for the three types of lubri-cooling conditions considered.

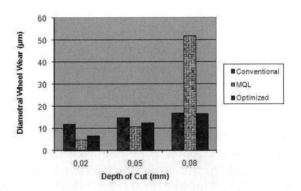

Figure 4.26. *Grinding wheel wear*

Figure 4.26 illustrates that the diametric wheel wear obtained by MQL technique was lower at 0.02 and 0.05 mm depths of cut, whereas at 0.08 mm, the value obtained was much higher than those obtained by the two other methods.

These behaviors can also be observed with the cutting force data, since the higher the wheel strength, the higher the cutting forces needed and, consequently, the more accentuated the tool wear.

Undoubtedly, the more the fluid penetrates the cutting zone, the less the wear is produced by the friction and the better is the removal of the generated heat. The porosity of the grinding wheel should be considered, since it is responsible for the accommodation of chips, thus improving the access of the fluid to the cutting zone, with consequent improvement in lubrication.

The vitrified bond is less susceptible to be eroded by the chips, presenting a greater retention force of the bond on the abrasive grain and, consequently, minimizing the diametric loss of the grinding wheel.

In the MQL method, at low depths of cut, parts of the chips mixed with the oil lodge in the wheel pores, allowing a larger proportion of the lubricant to reach the cutting zone. This also allows the airflow to remove part of the heat produced, improving the cooling of the place. So, even with this lodging that produces higher surface roughness, it contributes to a smaller wear of the tool.

However, when the machined chips become bigger (with increase in the depth of cut), the pores are not able to lodge them and the airflow is not efficient to remove them from the cutting zone. Thus, these damage the cutting operation and increase the abrasive tool strength, decreasing the lubrication capacity, thus generating surface damage and more accentuated wheel wear.

The optimized method also presented satisfactory results, being much more efficient than those by the conventional method. The high speed of fluid application makes the chip and heat removal easier than that when abundant fluid flow is used. It justifies the reason why the optimized method showed better results, thus guaranteeing a longer wheel life.

The G-ratio is the ratio between the quantities of removed material and the quantities of worn material (wheel). This relationship was obtained from the grinding wheel wear.

Figure 4.27 shows that MQL is a good alternative for low depths of cut. The optimized method needs to be highlighted, since it shows that the high-speed fluid application is more efficient than that by the excessive fluid application in the conventional method. As the depth of cut increases, G-ratio values decrease because of the increase in tool wear. For medium values, the MQL and optimized methods provided the best results. In this condition, however, the MQL method presented high values for surface roughness, which is unsatisfactory concerning the surface quality required.

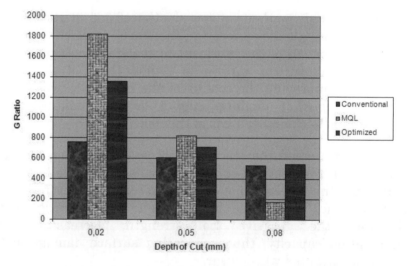

Figure 4.27. *G-ratio for each lubri-cooling method analyzed*

For high depths of cut, the MQL method was ineffective whereas the performances of conventional and optimized methods were similar because of the high quantity of heat to be removed even with the high-speed fluid application.

4.3.3.1.4. Microhardness

Microhardness in the ground workpieces was also analyzed. The graph in Figure 4.28 shows the gathered data and the reference value adopted, which represent the microhardness before grinding.

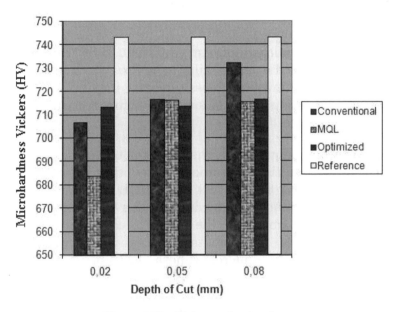

Figure 4.28. *Vickers microhardness*

The values obtained in the workpieces show that in all machining conditions and using the three methods of lubri-cooling resulted in a loss of surface microhardness. This is associated with the capacity of heat dissipation and cooling speed of the specimens, being the determinant of such factors to define the granular structure obtained.

Analyzing first the data gathered with 0.02 mm depth of cut, the loss obtained using the conventional method is smaller than that obtained by the MQL method, but higher than that obtained by the optimized method. This is consistent with the fact that MQL produces more heat during cutting, whereas the excessive and directed fluid jets of the conventional and optimized methods, respectively, remove more heat from the cutting zone. The ground workpieces under conventional lubri-cooling lost around 5% of their surface microhardness. Those ground by the optimized method lost around 4%, whereas those by MQL approximately 8%.

At 0.05 mm depth of cut, different fluid application methods presented a very similar behavior, which makes any conclusion about microhardness tendency impossible. On the other hand, there is a loss of about 4% compared with a non-ground workpiece.

At 0.08 mm depth of cut, the values obtained by MQL are much lower than those obtained by the other methods. This can be explained by the fact that MQL in this condition produces too much heat during the cutting process, consequently damaging the workpiece (high surface roughness values) and the tool (excessive wear). In this grinding condition, microhardness loss of the ground workpiece by MQL was around 4% whereas it was only 1.5% by the conventional method. The optimized method was not very efficient in removing heat at high depths of cut, presenting a behavior similar to the MQL method and lower than that of the conventional method.

Another interesting fact observed in Figure 4.28 is that the microhardness loss in the most severe condition was smaller than that obtained in the mildest condition. One of the reasons is that the latter was exposed three times longer than the former. Therefore, there was more exposure to constantly generated heat, although this heat was also present in smaller quantities.

Thus, it can be concluded that the heat generated during grinding causes a reduction of the workpiece hardness, independently of the lubri-cooling method.

4.3.3.2. *Grinding of advanced ceramics with diamond wheels*

The methodology consisted of wheel preparation, process parameter verification, experimentation, and material characterization. Because of the relevance of wheel conditioning on the results [TON 96], each new experiment was preceded by profile dressing and exposition of the abrasives, following a standard procedure.

Experiments consisted of collecting data during five grinding cycles on the workpieces, quantifying the initial condition of grinding. After that, another specimen was ground, having 17 mm (16,000 mm^3) of its length removed, thus aiming to promote enough wear in order to measure the G-ratio.

The experiments were performed on a SulMecânica 1055E surface grinder. Surface roughness was measured by a Taylor Hobson Surtronic 3+ rugosimeter, adjusted for a 0.8-mm cut-off length.

The equipment used to apply the MQL technique was an Accu-Lube microlubrication system (ITW Chemical Products Ltd.), which works based on a pulsating method of providing oil through a stream of compressed air. In addition, it allows the flow adjustment of compressed air and lubricant oil to be performed separately.

Workpieces of 99.8% alumina were fabricated by uni-axial pressing at 600 MPa, followed by sintering at 1,600°C for 2 h, charred alumina A1000-SG (Almatis, Inc., Frankfurt, Germany) with particle equivalent average diameter of 0.4 μm, surface area of 7.7 m^2/g, and ρ_{real}: 3.99 g/cm^3.

The diamond grinding wheel (Nikon) is specified by D107N115C50, having 350-mm diameter, 15-mm thickness and being dressed (multiple-point dresser) and trued up (alumina rod of 320 grids) before each test.

The fluid used in the conventional condition was synthetic soluble oil by Mobil, which recommends a concentration between 4% and 6% and pH between 8.5 and 9.5.

Two different methods of lubri-cooling, with distinct nozzles, were used in this study. The conventional lubri-cooling method, which is characterized by the application of cutting fluid at a high flow rate and low pressure. The nozzle worked with a 27.5 L/min flow, pressure less than 0.2 kgf/cm², and outlet fluid speed of 3 m/s.

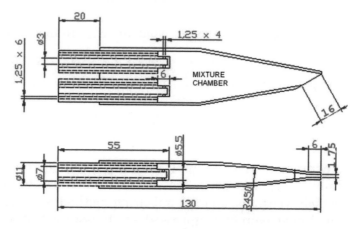

Figure 4.29. *MQL nozzle design*

To perform the MQL technique, a specific nozzle was used (Figure 4.29), which was designed to provide more efficiency in the cutting area. Airflow was determined assuming a pressure level of 8 kgf/cm²; the compressed air speed was the same as the wheel peripheral speed. The airflow used was 26.6 m³/h (450 L/min), whereas the oil flow determined during preliminary experiment was 80 mL/h (0.0013 L/min).

Two specimen geometries were defined, one to measure G-ratio and the other, 8 mm thick, 60 mm wide, and 120 mm long, to evaluate the damages as well as to measure tangential and normal forces, specific energy, acoustic emission, surface roughness, and other output variables.

4.3.3.2.1. Surface roughness

Surface roughness data obtained by MQL were higher than those obtained by the conventional method (Figure 4.30). Because of the fragile removal mode and intrinsic porosity of ceramic materials, the measurements presented a significant standard deviation. Moreover, although the grinding wheel was dynamically balanced, some vibration and its rigidity could have contributed to this variation.

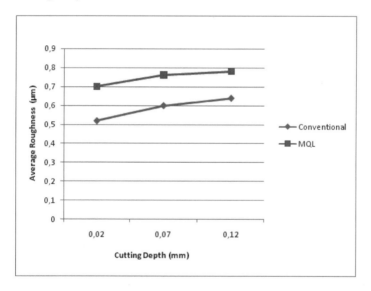

Figure 4.30. *Average surface roughnesses at different depths of cut*

The tendency for better finishing in the experiments through conventional lubri-cooling can be explained by a more efficient lubrication in the cutting zone, which, however, did not reflect a reduction in cutting forces because the excessive lubrication action was located in the central region of the workpiece. It is important to emphasize that all measurements were made along the specimen, perpendicular to the direction of the cut.

4.3.3.2.2. Diametral wheel wear

Grinding wheel wear data showed that MQL produced a smaller diametral wear than that by the conventional lubrication method. The diametral wear is represented in Figure 4.31.

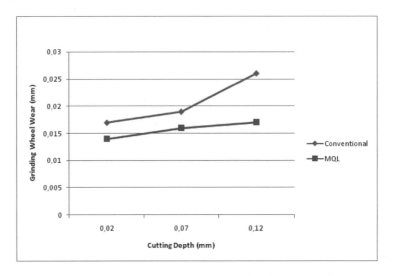

Figure 4.31. *Diametral wheel wear*

4.3.3.2.3. G-ratio

The experiments carried out by the conventional method presented lower G-ratio than those by the MQL method. The influence of the depth of cut was also observed. The graph showing the G-ratio is represented in Figure 4.32.

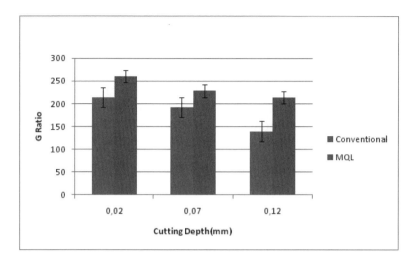

Figure 4.32. *G-ratio for each lubri-cooling condition*

Higher depths of cut lead to lower values of G-ratio, which indicate that the main wear mechanism is the abrasion microwear and the acting forces in grinding are not enough to promote microfracture and removal of abrasive grains.

4.4. Conclusions

As can be seen, MQL has been widely applied in different machining processes, including grinding (and its different types). The need of sustainable manufacture stimulated the research and study of alternative lubri-cooling methods, such as the optimized method and the MQL technique.

Nevertheless, if carefully applied in grinding, MQL can provide satisfactory results regarding surface quality and microstructural integrity of the workpiece. Moreover, it results in environmentally friendly and technologically relevant gains.

The door is open for future research in this field for the optimization of nozzles, cutting fluid composition, and the control of machining parameters, along with some computational modeling and simulation concerning thermal distribution and fluid flow.

In addition, cost estimations should be done for each case to enable more efficient applications of MQL. The costs related to this technology can be offset by the lack of the need for maintenance and disposal of cutting fluid, which today represents a considerable cost because of the current standards aiming at environmental preservation.

4.5. Acknowledgments

The authors are indebted to FAPESP and CNPq (Brazil) for its financial support of the researches, to Elsevier and ABCM for granting permission for the re-use of the published materials, and to the Laboratory of Abrasive Machining (LUA) and the Laboratory of Data Acquisition and Signal Processing (LADAPS) for ensuring the success of these experiments.

4.6. References

[BAH 98] BAHETI U., GUO C., MALKIN S., "Environmentally-conscious cooling and lubrication for grinding", in *Proceedings of the International Seminar on Improving Machine Tool Performance*, San Sebastián, Spain, vol. 2, p. 643-654, 1998.

[HAF 01] HAFENBRAEDL D., MALKIN S., "Technology environmentally correct for internal cylindrical grinding", *Machines and Metals Magazine*, vol. 426, p. 40-55, 2001.

[HEI 98] HEISEL U., LUTZ D., WASSMER R., WALTER U., "The minimum quantity lubricant technique and its application in cutting processes", *Machines and Metals Magazine*, vol. 386, p. 22-38, 1998.

[INA 91] INASAKI I., "Monitoring and optimization of internal grinding process", *CIRP Annals – Manufacturing Technology*, vol. 40, no. 1, p. 359-362, 1991.

[KLO 00] KLOCKE F., BECK T., EISENBLÄTTER O., FRITSCH R., LUNG D., PÖHLS M., "Applications of minimal quantity lubrication (MQL) in cutting and grinding", in *Proceedings of the 12th International Colloquium Tribology Industrial and Automotive Lubrication*, Technische Akademie Esslingen, Ostfildern, Germany, 2000.

[KLO 97] KLOCKE F., EISENBLÄTTER G., "Dry cutting", *CIRP Annals – Manufacturing Technology*, vol. 46, no. 2, p. 519-526, 1997.

[LEE 01] LEE E.S., KIM N.H., "A study on the machining characteristics in the external plunge grinding using the current signal of the spindle motor", *International Journal of Machine Tools and Manufacture*, vol. 41, issue 7, May 2001, pp. 937-951.

[LIA 00] LIAO Y.S., LUO S.Y., YANG T.H., "A thermal model of the wet grinding process", *Journal of Materials Processing Technology*, vol. 101, nos. 1-3, p. 137-145, 2000.

[MAL 89] MALKIN S., *Grinding Technology: Theory and Application of Machining with Abrasives*, Ellis Horwood Ltd., Chichester, England, 1989.

[SIL 05] SILVA L.R., BIANCHI E.C., CATAI R.E., FUSSE R.Y., FRANÇA T.V., AGUIAR P.R., "Study on the behavior of the minimum quantity lubricant – MQL technique under different lubrication and cooling conditions when grinding ABNT 4340 steel", *Journal of the Brazilian Society of Mechanical Sciences and Engineering*, vol. 27, pp. 192-199, 2005.

[SIL 07] SILVA L.R., BIANCHI E.C., FUSSE R.Y., CATAI R.E., FRANÇA T.V., AGUIAR P.R., "Analysis of surface integrity for minimum quantity lubricant – MQL in grinding", *International Journal of Machine Tools and Manufacture*, vol. 47, no. 2, p. 412-418, 2007.

[TAW 03] TAWAKOLI T., "Minimum coolant lubrication in grinding", *Industrial Diamond Review*, IDR 1, 2003, p. 60-65.

[TAW 07] TAWALOKI T., WESTKÄMPER E., RABIEY M., RASIFARD A., "Influence of the type of coolant lubricant in grinding with CBN tools", *International Journal of Machine Tools and Manufacture*, vol. 45, p. 734-739, 2007.

[TAW 07a] TAWAKOLI T., WESTKÄMPER E., RABIEY M., "Dry grinding by special conditioning", *International Journal of Advanced Manufacturing Technology*, vol. 33, p. 419-424, 2007.

[TAW 08] TAWAKOLI T., AZARHOUSHANG B., "Influence of ultrasonic vibrations on dry grinding of soft steel", *International Journal of Machine Tools and Manufacture*, vol. 48, no. 14, p. 1585-1591, 2008.

[TAW 09] TAWAKOLI T., HADAD M.J., SADEGHI, M.H., DANESHI, A., STÖCKERT S., RASIFARD A., "An experimental investigation of the effects of workpiece and grinding parameters on minimum quantity lubrication – MQL grinding", *International Journal of Machine Tools and Manufacture*, vol. 49, nos. 12-13, p. 924-932, 2009.

[TON 96] TÖNSHOFF H.K., MEYER T., WOBKER H.G., "Machining advanced ceramics with speed-stroke grinding", *Ceramic Industry*, vol. 7, p. 17-21, 1996.

[WEB 95] WEBSTER, J., "Selection of coolant type and application technique in grinding", in *Conference Paper of Supergrind '95* "Developments in Grinding", pp. 205–220, Storrs, Connecticut, USA, November 2–3, 1995.

[WEB 95a] WEBSTER J., CUI C., MINDEK R.B. Jr., LINDSAY R., "Grinding fluid application system design", *CIRP Annals – Manufacturing Technology*, vol. 44, no. 1, p. 333-338, 1995.

[WEB 99] WEBSTER J.A., "Optimizing coolant application systems for high productivity grinding", *Abrasives Magazine*, Oct/Nov, p. 34-41, 1999.

Chapter 5

Single-Point Incremental Forming

This chapter begins with a state-of-the-art review of the incremental sheet metal forming (ISMF) processes and continues with the presentation of a new theoretical framework that is capable of dealing with the fundamentals of single-point incremental forming (SPIF). The framework accounts for the influence of major process parameters and their mutual interaction and was developed in the light of membrane analysis with bi-directional, in-plane contact friction forces and ductile fracture mechanics. The model is validated by means of experimentation and finite element (FE) analysis, and the estimates of the formability limits in the principal strain space are used to successfully address the production and industrial applications in the last part of the chapter.

5.1. Introduction

SPIF is a new sheet metal forming process with a high, potential economic payoff for rapid prototyping applications

---

Chapter written by Maria Beatriz SILVA, Niels BAY and Paulo A.F. MARTINS.

and for small-scale production. SPIF is an environmentally conscious forming process because it allows significant savings in both material and energy requirements that are needed for the production of tooling when compared with alternative manufacturing processes. In fact, the flexible and dieless characteristics of SPIF enable the production of different parts with the same tooling apparatus, reducing waste, and minimizing raw material usage.

5.2. Incremental sheet forming processes

The idea of employing incremental forming to produce sheet metal parts was patented by Leszak [LES 67] and consisted of an extension of the conventional spinning process. SPIF was developed from the incremental forming concept and was proven to be feasible by Kitazawa *et al.* [KIT 96] in manufacturing rotational symmetric parts of aluminum with a special-purpose CNC machine tool. The capability study of using an ordinary CNC milling machine, instead of a special-purpose CNC machine tool, which was later performed by Jeswiet [JES 01] and Micari *et al.* [FIL 02], was the starting point for the successful and rapid development of the process.

ISMF can be divided into three different categories [JES 05]: (i) SPIF; (ii) incremental forming with counter tool (IFCT); and (iii) two-point incremental forming (TPIF). These categories are discussed in the following sections.

5.2.1. *Single-point incremental forming*

In SPIF, a sheet is clamped rigidly around its edges, but unsupported underneath, and formed with a spherically ended forming tool, which describes the contour of the desired geometry. The process can be performed with or without a backing plate [JES 05] but the use of a backing plate helps define the bottom contour of the geometry

(Figure 5.1). SPIF is the most flexible approach of ISMF processes.

Figure 5.1 presents the basic components of the process: (i) sheet metal blank; (ii) blankholder; (iii) backing plate; and (iv) rotating single-point forming tool.

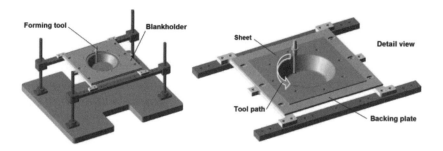

Figure 5.1. *Schematic representation of the SPIF process*

The tool path is generated in a computer-aided manufacturing (CAM) program and is used to progressively form the sheet metal blank into a component. During the process, there is no backup die supporting the back surface of the sheet.

The main advantages of SPIF are as follows [JAD 03]:

– sheet metal parts produced directly from CAD data;

– no need for a positive or a negative shaping die;

– easy and fast design changes, such as in machining;

– large increase in material formability;

– no need for special machine; conventional CNC milling machine can be used;

– small forces due to incremental nature of process;

– part size limited only by the machine tool; and

– good surface finish of formed parts.

The main disadvantages of the SPIF process are as follows:

– much longer forming time than in conventional stamping;

– process applicability limited to small-scale production;

– formation of right angles possible only by multi-step strategies;

– final accuracy of formed parts low because of springback. Correction algorithms must be used to minimize errors in accuracy; and

– geometric accuracy is lower than in other ISMF processes, especially in convex radii and bending edges [HIR 06].

Despite the major contributions of several researchers to the development of industrial applications and better characterization of the forming limits of the process [JES 05], the deformation mechanics of SPIF remains little understood. To fill the aforementioned gap in knowledge, this chapter presents a closed-form analytical framework that allows the influence of major fundamental process parameters and their mutual interaction to be studied both qualitatively and quantitatively. The analytical framework is successful in explaining the experimental and numerical results that were made available in the literature during the past couple of years.

5.2.2. *Incremental forming with counter tool*

An IFCT is a variant of SPIF without a backing plate (Figure 5.2). The counter tool makes a trajectory similar to the trajectory of the main forming tool during which the fixture moves downward (Figure 5.2).

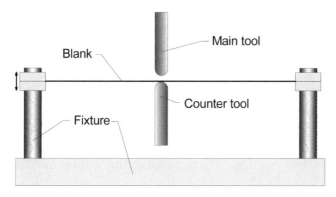

Figure 5.2. *Schematic representation of the IFCT process*

If the overall shape of the sheet metal part is complex, the trajectory of the counter tool needs to be different from that of the main tool. Because the movement of the counter tool requires independent control, a special-purpose machine or an auxiliary driving mechanism is required.

5.2.3. *Two-point incremental forming*

In TPIF, the sheet metal blank is formed against partial or full dies with one or more spherically-ended forming tools (Figure 5.3).

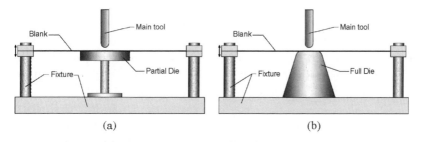

Figure 5.3. *Schematic representation of the TPIF process: (a) TPIF partial die; (b) TPIF full die*

The sheet is clamped rigidly around its edges with a blankholder that moves vertically. The forming tool moves along a trajectory on the outer surface of the metal sheet part, from the top to the bottom of the geometry.

5.2.3.1. *TPIF with a partial die*

The partial die in TPIF is comparable with a backing plate in SPIF and supports only essential areas of the blank. The partial die has unspecific geometry, enabling it to form different but similar shapes with the same die [JES 05, MAT 01].

The geometric accuracy of a sheet metal part obtained by TPIF is better than that of a similar part obtained by SPIF.

5.2.3.2. *TPIF with a full die*

The TPIF with a full die is not considered a dieless approach. However, the process is directly related to other ISMF processes and successfully used in the production of prototypes.

Comparing the TPIF with a full die to all the other dieless approaches, it can be concluded that the process is capable of ensuring good geometric accuracies of the final parts [HIR 06, JES 05]. This is because the sheet metal blank is constrained by both the tool and the die during forming.

In terms of costs, TPIF with a full die is the most expensive incremental forming process because of the extra costs associated with design and manufacturing of the die. The die can be fabricated in several materials such as steel, aluminum, plastic, wood, and foam blocks [JES 05].

The main disadvantage of the process is its poor flexibility because a new die is needed for the production of each new part, as in conventional sheet metal forming.

5.3. Analytical framework

In past years, most studies on SPIF concerned experimental investigations on applications and formability limits, as well as that a limited number of FE studies have been carried out. The keynote paper by Jeswiet *et al.* [JES 05] presents a comprehensive state-of-the-art review and includes a thorough list of references on the most significant, published research work in the field. Results from FE investigations were used to address the mechanics of deformation, and failure has been related to both the sine law and the spinnability relation given by Kegg [KEG 61] (i.e. emphasizing on the importance of axis-parallel shear as in the case of shear spinning).

More recently, the governing mode of deformation in SPIF has been a subject of controversy in the metal forming community [FIL 02]. Some authors claim that deformation takes place by stretching instead of shearing, whereas others claim the opposite, but assertions are mainly based on "similarities" with well-known processes of stamping and shear spinning rather than experimental evidence from SPIF itself. However, as recently shown by the authors [SIL 08], the examination of the likely mode of material failure at the transition zone between the inclined wall and the corner radius of the sheet is consistent with stretching, rather than shearing, being the governing mode of deformation in SPIF.

There are two different views on the formability limits of SPIF. The commonly accepted view considers (i) that formability is limited by necking; (ii) that the forming limit curve (FLC) in SPIF is significantly raised against conventional FLCs being used in the analysis of sheet metal forming processes (e.g. stamping and deep drawing, among others) [FIL 02]; and (iii) that the increase in formability is either due to a large amount of through-thickness shear [ALL 07] or due to serrated strain paths arising from cyclic, local plastic deformation [EYC 07]. This approach (which hereafter is referred to as the "necking line of attack") is adopted in most of the numerical and experimental contributions to the understanding of formability in SPIF that were published in the past years.

The alternative (non-traditional) view of formability in SPIF, recently proposed by the authors [SIL 08] and supported by Cao *et al.* [CAO 08] and Emmens and Boogaard [EMM 08], considers (i) that formability is limited by fracture without experimental evidence of previous necking; (ii) that the suppression of necking in conjunction with the low growth rate of accumulated damage is the key mechanism for ensuring the high level of formability in SPIF; and (iii) that FLCs, which give the loci of necking strains, are not relevant and should be replaced by the fracture forming line (FFL). This approach hereafter is referred to as the "fracture line of attack".

As will be shown later in this chapter, when suppression of necking is included in the analysis, many shortcomings of the commonly accepted view on deformation and failure are removed. In particular, the experimental evidence that formability limits can be approximated, in the principal strain space, by lines of the form $\varepsilon_1 + \varepsilon_2 = q$ placed well above the conventional FLCs is something difficult to explain on the assumption that failure is limited by previous necking. In fact, if through-thickness shear or serrated strain paths arising from cyclic, local plastic deformation could increase the forming limits of aluminum AA1050-O to a level that is approximately six times that experimentally found by means of tensile, elliptical, and circular bulge tests [SKJ 09], then the individual effect of stresses and strain paths of SPIF on the FLCs would be much larger than what is commonly seen in conventional sheet metal forming processes.

Moreover, recent experimental results obtained by the authors comparing SPIF, using forced tool rotation, with SPIF, using free tool rotation, which most of the time leads to no rotation, showed no influence on the overall formability. This observation draws conclusion that the influence of circumferential friction resulting from the contact between the tool and the sheet is negligible, a result that is in close agreement with previous assumptions of the authors concerning frictional effects [SIL 08]. This result also seems to indicate that the level of through-thickness shear arising from the contact with friction between a single-point forming tool and a sheet having small thickness is not very significant.

The forthcoming sections are directed at presenting the analytical framework of SPIF that is based on membrane analysis with in-plane contact friction forces and discussing the formability limits derived from ductile damage mechanics in the light of experimental measurements and numerical results provided by FE analysis. Results show that there is good agreement between analytical, FE, and experimental results, implying that the proposed analytical framework can be easily and effectively used for explaining the mechanics of deformation and the formability limits of SPIF.

5.3.1. *Membrane analysis*

Circle grid analysis in conjunction with the observation of smear-mark interference between the tool and the sheet surface allows the classification of all possible tool paths as combinations of the basic modes of deformation depicted in Figure 5.4: (A) flat surfaces under plane strain stretching conditions; (B) rotational symmetric surfaces under plane strain stretching conditions; and (C) corners under equal bi-axial stretching conditions. It is worth noting that in between these modes of deformation, there are other possibilities in which neither plane strain stretching nor equal bi-axial stretching appear.

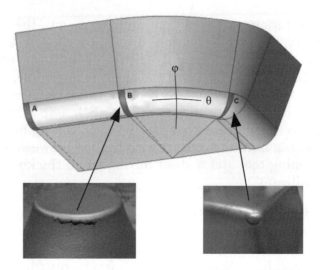

Figure 5.4. *Schematic representation of the volume element resulting from the local contact between the tool and the sheet placed immediately ahead of tool and identification of the basic modes of deformation of SPIF: (A) flat surface under plane strain stretching conditions; (B) rotational symmetric surface under plane strain stretching conditions, and (C) corner under bi-axial stretching conditions (note: inserts show images of typical cracks occurring in deformation modes B and C)*

The analytical framework of SPIF is focused on the aforementioned extreme modes of deformation that are likely to be found in SPIF. This model was recently published by the authors [SIL 08].

5.3.2. *State of stress and strain*

In SPIF, the local shell element CDEF in Figure 5.5 is subjected to normal forces, shear forces, and bending moments so that it conforms to the hemi-spherical shape of the tip of the pin tool, forming a contact area (A, B or C in Figure 5.4) between the tool and the part of the sheet placed immediately ahead of the moving tool.

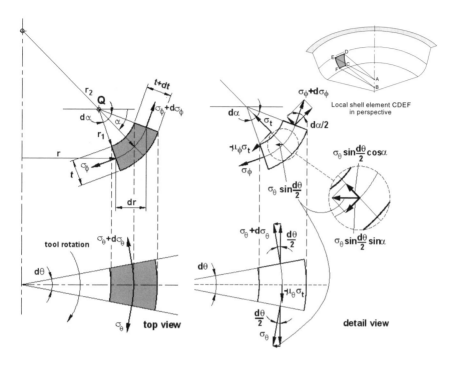

Figure 5.5. *Schematic representation of the shell element used in the membrane analysis of SPIF*

The state of strain and stress acting in these areas can be derived from the membrane equilibrium conditions if bending moments are neglected and circumferential, meridional, and thickness stresses are assumed to be principal stresses. Further simplifying assumptions are as follows: the material is assumed to be rigid, perfectly plastic and isotropic and the resultant friction stress acting in the tool-sheet contact interface is assumed to consist of two in-plane components: a meridional component, $-\mu_\phi \sigma_t$, due to the downward movement of the tool, and a circumferential component, $-\mu_\theta \sigma_t$, due to the circumferential feed combined with the rotation of the tool. This last assumption, which is an untraditional way of modeling friction introduced for convenience, implies that the coefficient of friction can be given by equation [5.1].

$$\mu = \sqrt{\mu_\phi^2 + \mu_\theta^2} \qquad [5.1]$$

Resolving the force equilibrium along the thickness, circumferential, and meridional directions results in:

$$\sigma_t r \, d\theta \, r_1 \, d\alpha + \sigma_\phi r \, d\theta \, t \sin\frac{d\alpha}{2} + (\sigma_\phi + d\sigma_\phi)(r + dr) d\theta (t + dt) \sin\frac{d\alpha}{2} +$$
$$\sigma_\theta r_1 \, d\alpha \, t \sin\frac{d\theta}{2} \cos\alpha + (\sigma_\theta + d\sigma_\theta) r_1 \, d\alpha \, t \sin\frac{d\theta}{2} \cos\alpha = 0$$

$$\sigma_\theta r_1 \, d\alpha \left(t + \frac{dt}{2}\right) - \mu_\theta \sigma_t r_1 \, d\alpha \left(r + \frac{dr}{2}\right) d\theta - (\sigma_\theta + d\sigma_\theta) r_1 d\alpha \left(t + \frac{dt}{2}\right) = 0 \quad [5.2]$$

$$(\sigma_\phi + d\sigma_\phi)(r + dr) d\theta (t + dt) - \sigma_\phi r \, d\theta \, t + \mu_\phi \sigma_t r \, d\theta \, r_1 \, d\alpha - \sigma_\theta \frac{d\theta}{2} r_1 \, d\alpha \, t \sin\alpha -$$
$$(\sigma_\theta + d\sigma_\theta) \frac{d\theta}{2} r_1 \, d\alpha \, t \sin\alpha = 0$$

The distribution of stresses in the small localized plastic zones (A, B and C), which, as mentioned earlier, are the three typical deformation zones in SPIF, can easily be obtained from equation [5.2] after neglecting higher order terms, taking into account geometrical simplifications, considering the Tresca yield criterion and assuming that SPIF of flat and rotationally symmetric surfaces is performed under plane strain conditions, $d\varepsilon_\theta = 0$ [SIL 08].

Table 5.1 provides the strains and stresses along the principal directions that are derived from the analytical framework of SPIF. As mentioned previously, the model assumes flat and rotationally symmetric surfaces (A and B in Figure 5.4) to be formed under plane strain conditions, $d\varepsilon_\theta = 0$, and corners (C in Figure 5.4) to be formed under equal bi-axial stretching, $d\varepsilon_\phi = d\varepsilon_\theta > 0$. The inclined wall of the sheet adjacent to the forming tool is loaded under elastic uni-axial tension conditions. Further details with complete derivation of the equations can be obtained from the work of Silva et al. [SIL 08].

	Assumption	State of strain	State of stress	Hydrostatic stress
SPIF (flat and rotational symmetric surfaces)	Plane strain conditions	$d\varepsilon_\phi = -d\varepsilon_t > 0$ $d\varepsilon_\theta = 0$ $d\varepsilon_t < 0$	$\sigma_\phi = \sigma_1 = \dfrac{\sigma_Y}{(1+t/r_{tool})} > 0$ $\sigma_\theta = \sigma_2 = \dfrac{1}{2}(\sigma_1 + \sigma_3)$ $\sigma_t = \sigma_3 = -\sigma_Y \dfrac{t}{(r_{tool}+t)} < 0$	$\sigma_m = \dfrac{\sigma_Y}{2}\left[\dfrac{r_{tool}-t}{r_{tool}+t}\right]$
SPIF (corners)	Equal bi-axial stretching	$d\varepsilon_\phi = d\varepsilon_\theta > 0$ $d\varepsilon_t < 0$	$\sigma_\phi = \sigma_\theta = \sigma_1 = \dfrac{\sigma_Y}{(1+2t/r_{tool})} > 0$ $\sigma_t = \sigma_3 = -2\sigma_Y \dfrac{t}{(r_{tool}+2t)} < 0$	$\sigma_m = \dfrac{2\sigma_Y}{3}\left[\dfrac{r_{tool}-t}{r_{tool}+2t}\right]$
SPIF (inclined walls)	Uni-axial tension	$d\varepsilon_\phi > 0$ $d\varepsilon_\theta = d\varepsilon_t < 0$	$\sigma_{\phi D} = \sigma_{\phi C}\dfrac{r_C}{r_D} < \sigma_Y$ $\sigma_\theta = \sigma_t = 0$	$\sigma_m = \dfrac{\sigma_\phi}{3}$

Table 5.1. *State of stress and strain in the small localized plastic zones (A, B and C) and inclined walls of SPIF*

5.3.3. *Formability limits*

The study of the morphology of the cracks together with the measurement of thickness along the cross-section of SPIF parts revealed that plastic deformation takes place by uniform thinning until fracture without experimental evidence of localized necking takes place before reaching the onset of fracture [SIL 08].

The suppression of localized necking in SPIF is due to the inability of necks to grow. If a neck were to form at the small plastic deformation zone in contact with the incremental forming tool, it would have to grow around the circumferential bend path that circumvents the tool. This is difficult and creates problems of neck development. Even if the conditions for localized necking could be met at the small plastic deformation zone in contact with the tool, growth would be inhibited by the surrounding material that experiences considerably lower stresses. This implies that FLCs of conventional sheet metal forming are inapplicable to describe failure in SPIF. Instead, FFLs showing the fracture strains, placed well above the FLCs, should be employed.

The fracture forming limits in SPIF can be characterized by means of ductile damage mechanics based on void growth models. Assuming the Tresca yield criterion, linear loading paths, and that the damage function $f(\sigma_m/\bar{\sigma})$ takes the simple form of the tri-axiality ratio $\sigma_m/\bar{\sigma}$ [AYA 84], the total amount of accumulated damage for plane strain and equal bi-axial stretching SPIF conditions results in the following critical damage values:

$$D_c = \int_0^{\bar{\varepsilon}_f} \frac{\sigma_m}{\bar{\sigma}} d\bar{\varepsilon} = \frac{1}{2}\left[\frac{r_{tool}-t}{r_{tool}+t}\right]\varepsilon_1^{plane\,strain}$$

$$D_c = \int_0^{\bar{\varepsilon}_f} \frac{\sigma_m}{\bar{\sigma}} d\bar{\varepsilon} = \frac{2}{3}\left[\frac{r_{tool}-t}{r_{tool}+2t}\right]2\varepsilon_1^{bi-axial}$$

[5.3]

If the critical value of damage D_c at the onset of cracking is assumed to be path independent, solving equation [5.3] for ε_1 results in the following identity:

$$\frac{\varepsilon_1^{bi-axial}-\varepsilon_1^{plane\,strain}}{\varepsilon_2^{bi-axial}-0} = \frac{\frac{3}{4}\left[\frac{r_{tool}+2t}{r_{tool}-t}\right]-2\left[\frac{r_{tool}+t}{r_{tool}-t}\right]}{\frac{3}{4}\left[\frac{r_{tool}+2t}{r_{tool}-t}\right]} = -\frac{5\left(\frac{r_{tool}}{t}\right)+2}{3\left(\frac{r_{tool}}{t}\right)+6}$$

[5.4]

Equation [5.4] gives the slope of the FFL in the principal strain space $(\varepsilon_1, \varepsilon_2)$. For typical experimental values of r_{tool}/t_1

in the range 3 to 50 (corresponding to r_{tool}/t_0 in the range 2 to 10 for the investigated material), the slope derived from equation [5.4] will vary between −1.1 and −1.5. This supports the assumption that the fracture forming limit in SPIF can be approximately expressed as $\varepsilon_1 + \varepsilon_2 = q$, where $\varepsilon_t = -q$ is the thickness strain at the onset of fracture under plane strain conditions and ε_1 and ε_2 are the major and the minor principal strains in the plane of the sheet, respectively.

As is shown later in section 4, this result is in close agreement with the typical loci of failure strains in conventional sheet forming processes, in which the slope of the FFL is often about −1 [ATK 97, EMB 81].

It is worth noticing that the proposal regarding the suppression of localized necking in SPIF proposed by the authors [SIL 08] is consistent with a recent published work by Emmens and Boogaard [EMM 08]. The main difference between the approaches is related to the physics behind the suppression of necking. The authors [SIL 08] claim that there is an inability of necks to grow around the circumferential bend path that circumvents the tool in a zone surrounded by material undergoing elastic deformation, whereas Emmens and Boogaard [EMM 08] claim that the suppression of necking is a consequence of repeated bending and unbending loading cycles arising from the tool path.

5.4. FE background

Validation of the analytical framework of SPIF by means of numerical modeling was performed using the commercial FE computer program LS-DYNA (version ls971s). LS-DYNA is based on an explicit dynamic elastoplastic formulation and has been reported by other researchers to be capable of taking into account the practical nonlinearities in the geometry and material properties that are typical of SPIF to produce good predictions of shape, strain, and thickness distributions throughout the SPIF parts [SKJ 09].

5.4.1. *Modeling conditions*

The FE model of the sheet blanks is built on an initial course grid of 26 × 26 shell elements, each having a side length of about 9.7 mm (type 16 in LS-DYNA) for a typical part size of 253 × 253 mm. Adaptive grid refinement is used throughout the calculation to limit the interference between the sheet and the contours of both the forming tool and the backing plate and to obtain high levels of accuracy in terms of geometry and distribution of field variables. The adaptive grid refinement procedure consists of three refinement operations ending up by splitting the original elements into 64 new elements that have one-eighth of the initial element size. This is known to generate incompatible grids that must be treated with multi-point constraints in the transition region. The grid refinement is done ahead of the tool to ensure a fully refined grid in the zone of plastic deformation. A full-integration scheme is used with five integration points over the sheet thickness.

The description of both the forming tool and the backing plate is obtained by means of surface grids. Both active tool components are considered rigid and a large number of elements are used for modeling their geometry to reduce the level of roughness that is artificially introduced by the overall discretization procedure. The movement of the tool in the FE model, including the rotation and the helical path, which is defined by means of a large number of discrete points, is identical to that in the actual SPIF process. The number of points is determined by the tolerance setting in the CAM program. A tolerance of 0.01 mm was used, resulting in about 400 points/m for the hyperbolic cone and 8 points/m (the sides are straight) for the pyramid. Vertical step-down was set at 0.5 mm and the coefficient of friction adopting the Amonton-Coulomb law was assumed to be zero ($\mu = 0$).

Acceleration of the overall CPU time was performed by means of a load-factoring (or time-scaling) procedure that changed the rate of loading by an artificial increase in the velocity of the single-point forming tool by a factor of 1,500 times as compared with the real forming velocity. The maximum increment in time step for performing the explicit (central difference) time integration scheme was based on a characteristic length equal to the shell area divided by the longest diagonal. As a precaution, LS-DYNA uses 0.9 times this value to guarantee stability. In practical terms, this results in an average time step per increment of approximately 2.7×10^{-7} s.

The material of the sheet was considered rigid perfectly plastic, $\sigma_Y = 100$ MPa, to allow comparison between the results predicted by LS-DYNA and those calculated by means of the analytical framework developed by the authors. No anisotropy effects were taken into consideration.

FE simulation of SPIF under these modeling conditions is computationally very intensive and therefore a full-scale model (i.e. a model that does not take advantage of the existing symmetric conditions of the benchmark SPIF parts under investigation) is required with 120 to 240 h of the CPU time in a 900-MHz computer or 30 to 60 h of the CPU time in a quad-core processor computer.

5.4.2. *Post-processing of results*

After finishing the FE simulations of SPIF, calculated values of thickness were taken from different sets of shell elements located in the small plastic contact zone between the single-point forming tool and the sheet surface.

The calculated values of thickness were used for determining the analytical distribution of the stress field from the equations listed in Table 5.1. Furthermore, values of the components of thickness, circumferential, and meridional strains and stresses calculated by FE analyses were taken from selected areas of the formed geometry for comparison with the results provided by the analytical model (Figure 5.6).

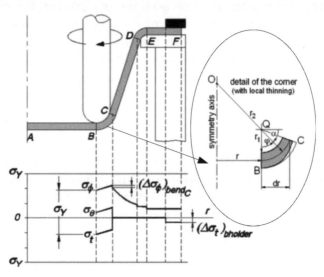

Figure 5.6. *Schematic representation of the stress field under plane strain conditions in the radial slice of the component being formed that contains the local shell element used for the membrane analysis of SPIF*

All values were obtained by averaging FE results at the integration points and at different, arbitrary locations corresponding to the contact areas A, B, and C in Figure 5.4 and the inclined wall surface of the sheet adjacent to the forming tool. The elements of each set are positioned along the meridional direction and are located at a safe distance from the backing plate (more than 10 times the initial sheet thickness apart from it).

5.5. Experimental

This section starts by describing the experimental techniques that were used for obtaining the forming limits and the fracture forming limits and continues by presenting the procedure that was used in the SPIF tests. All specimens were obtained from AA1050-H111 sheet blanks with 1-, 1.5-, and 2-mm thickness.

5.5.1. *Forming and fracture forming limit diagrams*

Formability of the sheet material was evaluated by means of tensile tests (using specimens cut at 0°, 45°, and 90° with respect to the rolling direction) and bi-axial circular (100 mm) and elliptical (100/63 mm) hydraulic bulge tests (Figure 5.7). The experimental technique used for obtaining the FLC involved electrochemical etching a grid of circles with 2-mm initial diameter on the surface of the sheets before forming and measuring the major and minor axes of the ellipses that result from the plastic deformation of the circles during the formability tests. The FLC was constructed by taking the strains $(\varepsilon_1, \varepsilon_2)$ at failure from grid elements placed just outside the neck (i.e. adjacent to the region of intense localization) since they represent the condition of the uniformly thinned sheet just before necking occurs. The experimental technique is described elsewhere [ROS 76] and the resulting FLC is plotted in Figure 5.7.

Although sheet metal thickness normally increases the FLCs (i.e. the thicker the sheet, the higher the FLC), it was decided to plot the curve shown in Figure 5.7 from the entire set of formability tests on the three different sheet thicknesses investigated. The arguments for doing this are that the necking strains turned out to be almost the same for specimens with 1-, 1.5-, and 2-mm thickness and that the FLC depicted in Figure 5.7 is to be used only for qualitative analysis in the forthcoming sections.

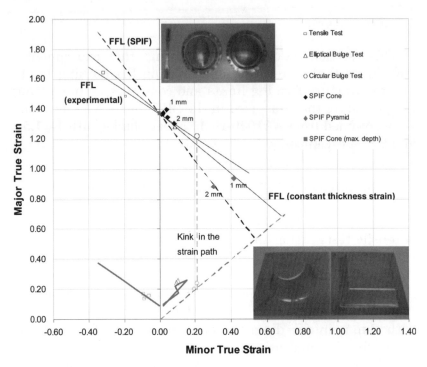

Figure 5.7. *Fracture forming limit diagram containing the FLC, the FFL, and the fracture points obtained from conical and pyramidal SPIF parts*

The intersection of the FLC with the major strain axis occurs at $\varepsilon_1 = 0.07$, which is in good agreement with the value of the strain-hardening exponent of the stress-strain curve obtained by means of tensile tests

$$\sigma = 140\,\varepsilon^{0.041}\ \text{MPa} \tag{5.5}$$

The experimental FFL is more difficult to construct than the FLC. Application of grid technique with very small circles to obtain strains in the necking region close to the fracture provides strain values in the direction perpendicular to the crack direction, which cannot be considered representative for the fracture strains. Moreover, such grids create measurement problems and suffer from sensitivity to the initial size of the circles used in the grid because of the inhomogenous deformation in the neighborhood of the crack.

Because of this, the two other principal strains were measured and the strain perpendicular to the crack was then determined from these strains by volume constancy. The strain parallel to the crack direction was determined by the grid technique, whereas the local thickness at fracture was measured with a microscope at several places along the crack to obtain the "gauge length" strains. This procedure, besides being time consuming, is increasingly difficult to execute as the thickness of the specimens becomes smaller.

In the present investigation, gauge length strains and the corresponding FFL were determined from the material tests and SPIF formability tests performed on specimens with 2-mm initial thickness excluding tests on the 1-mm sheet. This is acceptable because in the strict fracture mechanics sense, the position of the FFL in the principal strain space shifts upward or downward on the positive strain axis only as a function of the ductility of the material.

However, comparing this assumption and related experimental procedure with equation [5.4], we may question whether the FFL should be considered as a material-dependent line or a material- and process-dependent line. In the second case, the FFL would shift upward or downward depending on process variables such as the thickness and radius of tooling (equation [5.4]), which directly affect the level of accumulated damage. Yet, because the distance between the FLC and the FFL is very large, it is reasonable to consider that process variables have limited influence and that the FFL is a material-dependent line.

The experimental FFL is plotted in Figure 5.7 and can be approximated by a straight line ($\varepsilon_1 + 0.79\,\varepsilon_2 = 1.37$) falling from left to right, which is close to the condition of constant through-thickness strain at fracture (given by a slope of –1). The large distance between neck formation (FLC) and fracture (FFL) indicates that AA1050-H111 is a very ductile material that allows a considerable through-thickness strain between neck initiation and fracture.

Through-thickness straining in the principal strain space is characterized by a sharp bend in the strain path after local neck formation, as can be seen in the case of the formability tests that were used for constructing the FFL (Figure 5.7). The strain paths of bi-axial, circular and elliptical bulge formability tests show a kink after neck initiation toward vertical direction, that is, parallel to plane strain conditions, as schematically plotted by the gray dashed line corresponding to the circular bulge formability test. The strain paths of tensile formability tests also undergo a significant change of strain ratio from a slope of –2 to a steeper one, although not to vertical direction. The absence of a sharp kink of the strain path into vertical direction in tensile formability testing, instead of a less abrupt bend, is because major and minor strains after the onset of necking need to be taken into account from directions that are not coinciding with the original pulling direction. A comprehensive analysis on the direction of the strain paths in the tension-compression strain quadrant can be found in the work of Atkins [ATK 96].

5.5.2. *SPIF experiments*

The experiments were performed in a Cincinnati Milacron machining center equipped with a rig, a backing plate, a blankholder for clamping the sheet metal blanks, and a rotating, single-point forming tool (Figure 5.1). The forming tool with a diameter of ø 12 mm and a hemi-spherical tip was made of cold working tool steel (120WV4-DIN) hardened and tempered to 60 HRC in the working region. The speed of rotation was 35 rpm and the feed rate was 1,000 mm/min.

The tool path was helical, with a step size per revolution equal to 0.5 mm. The lubricant applied between the forming tool and the sheet was diluted cutting fluid.

The experiments were designed with the objective of measuring the strain values at fracture and the tests were performed in truncated conical and pyramidal shapes characterized by stepwise drawing angles ψ varying with the depth (Figure 5.8). A grid with 2-mm circles was electrochemically etched on the surface of the sheets, allowing the principal strains to be measured. The procedure considers the thickness strain ε_t to be a principle strain and assumes the sheet blank to be deformed by membrane forces so that it conforms to the shape of the tool path. The experiments were performed in random order and at least two replicates were produced for each combination of thickness and geometry to provide statistical meaning.

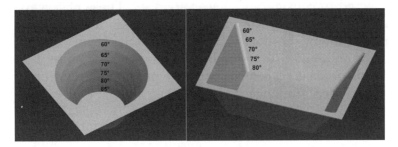

Figure 5.8. *Geometry of the truncated conical and pyramidal shapes that were used in the experiments*

5.6. *Results and discussion*

The first part of this section examines the mechanics of deformation of SPIF derived from the analytical framework and compares the theoretical distribution of strains and stresses with the numerical estimates provided by the FE analysis. The second part focuses on the assessment of the formability limits proposed by the analytical framework against experimental measurements of SPIF parts at failure.

5.6.1. *Stress and strain fields*

Figure 5.9 shows the analytical and FE estimates of the stress and strain fields in the small, localized plastic zone during SPIF of a truncated conical shape of a rigid perfectly plastic metal sheet (σ_Y = 100 MPa) with 1-mm initial thickness.

The distribution of the normalized effective stress calculated by the FE analysis (Figure 5.9a) confirms the incremental and localized characteristics of the deformation. Plastic deformation occurs only in the small radial slice of the component being formed under the tool. The surrounding material experiences elastic deformation and therefore is subjected to considerably lower stresses. These observations together with the progressive decrease in the applied stresses along the inclined wall of the SPIF part (from the transition point C to point D near the backing plate; Figure 5.6) are in good agreement with the predictions derived from the analytical framework (Table 5.1).

The comparison between the FE estimates and the analytical values of the meridional ($\sigma_1 = \sigma_\phi$) and the circumferential ($\sigma_2 = \sigma_\theta$) stresses in the plastic deformation zone in three different locations placed along the small plastic deformation region further corroborates the applicability of the analytical framework developed by the authors (Figure 5.9c).

Major differences between numerical and analytical estimates are found in the distribution of thickness stress ($\sigma_3 = \sigma_t$), which may have two likely explanations. First, membrane elements are known to experience difficulties when the amount of through-thickness stress resulting from contact pressure in small radii is significant. Second, the analytical model considers a smooth flat transition between the corner and the undeformed bottom region of the SPIF part whereas the experimental observations and the FE calculations show a small concave depression (i.e. a "circumferential valley") at the transition region.

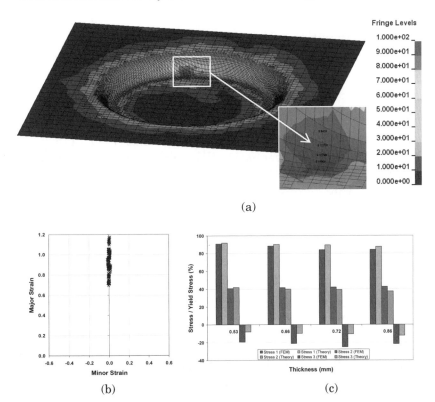

Figure 5.9. *SPIF of a truncated conical shape (material: rigid perfectly plastic; initial thickness: 1 mm): (a) FE distribution of the normalized effective stress (%), (b) FE distribution of major and minor true strains in the principal strain space, and (c) comparison between FE and theoretical estimates of the principal stresses for three different values of the final thickness of the cone corresponding to different locations in the plastic deformation zone*

The FE evolution of thickness along the small plastic deformation region placed under the forming tool shows that reduction in thickness (down to a value of $t = 0.45$ mm) tends to balance the increase in the meridional stress σ_ϕ. This result is in good agreement with the analytical framework that considers product $\sigma_\phi t$ to be constant in the plastic deformation region [SIL 08].

The distribution of major and minor true strains in the principal strain space obtained by means of the FE analysis confirms that SPIF of a truncated conical shape is performed under plane strain conditions because all strains occur close to the major strain axis (Figure 5.9b).

Results shown in Figures 5.10 and 5.11 also indicate a good agreement between the FE and analytical results for the SPIF of pyramidal shapes. It is worth noticing that the sides of the pyramids are formed under plane strain conditions (Figure 5.10b), whereas the corners are shaped with strain paths deviating toward equal bi-axial stretching conditions (Figure 5.11b). This last observation is important for the overall validation of the analytical framework because it confirms the extreme modes of deformation that were necessary to assume during the theoretical developments [SIL 08] (Figure 5.4).

However, a closer observation of the FE results shown in Figure 5.10b allows the conclusion that although bi-axial stretching occurs in the corners, it does not appear as balanced, equal, bi-axial deformation because of heavier constraints in tangential direction θ than in meridional direction ϕ. This is due to the variation in rigidity of the elastic region surrounding the local plastic deformation zone. In fact, the region above the deformation zone is fully deformed, thus having minimum thickness, whereas the elastic zone at the same depth as the deformation zone, but larger or smaller θ, has larger average thickness. In other words, the deviation from equal bi-axial into bi-axial stretching at the corners of the pyramid is probably because stiffness in the circumferential direction is much higher than in the meridional direction.

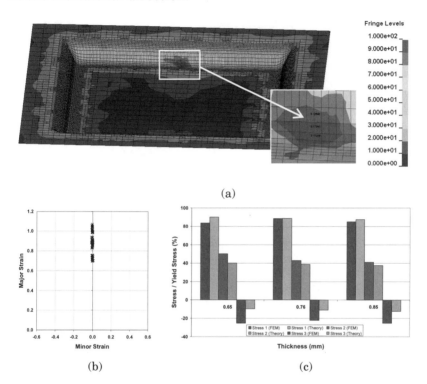

Figure 5.10. *SPIF of a pyramidal shape (material: rigid perfectly plastic; initial thickness: 1 mm; and zone: side of the pyramid): (a) FE distribution of the normalized effective stress (%), (b) FE distribution of major and minor true strains in the principal strain space, and (c) comparison between FE and theoretical estimates of the principal stresses for three different values of the final thickness of the cone corresponding to different locations in the plastic deformation zone*

5.6.2. *Formability limits*

The principal strain space in Figure 5.7 includes the strain signature's characteristic of necking as well as those of fracture. The concept of the FFL is that all possible fracture strains are located on a specific line, which characterizes the material in contrast to the concept of the FLC, which is sensitive to the strain path.

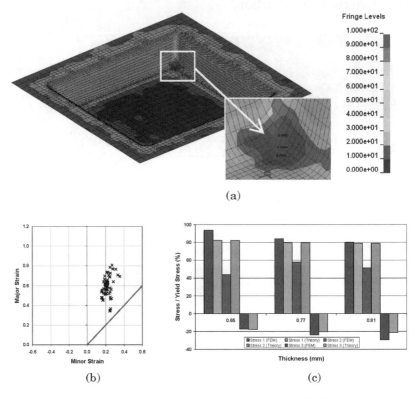

Figure 5.11. *SPIF of a pyramid shape (material: perfectly rigid plastic; initial thickness: 1 mm; and zone: corner of the pyramid): (a) FE distribution of the normalized effective stress (%); (b) FE distribution of major and minor true strains in the principal strain space; and (c) comparison between FE and theoretical estimates of the principal stresses for a specific location in the plastic deformation zone*

Figure 5.7 shows three different FFLs. The black, solid line, denoted as "FFL-experimental", is obtained from the fracture strains measured in the experimental tensile and bi-axial bulge tests. The black, thin, solid line, denoted as "FFL-constant thickness strain", is the theoretical fracture line falling from left to right in the strain space with a slope equal to −1. The black, dashed line, denoted as "FFL-SPIF", is the fracture line derived from the critical value of damage D_c at the onset of cracking for the analytical distribution of stresses and strains that was proposed by the authors (see equation [5.4]).

There is good agreement between these three FFLs and also between these lines and the experimental fracture strains measured for the conical and pyramidal shapes obtained from AA1050-H111 sheets of 1- and 2-mm initial thickness. The largest deviations adopt the "FFL-SPIF" line, which can be explained by the fact that this line was constructed under the assumption of a rigid perfectly plastic material.

It is worth noting that the experimental values of fracture strains for the pyramidal shapes are not placed on the equal bi-axial strain ratio line with a slope of −1 in the principal strain space. In fact, although the onset of failure is located at the corner of the pyramids, the values of fracture strains are somewhat deviating toward the plane strain direction. This verifies the earlier observation from the FE analysis concerning the occurrence of bi-axial deformation, instead of equal bi-axial deformation, due to heavier constraints in tangential direction θ than in meridional direction ϕ.

Finally, it is important to make reference to the solid gray squares placed exactly on top of both the major strain axis and the equal bi-axial strain direction. The fracture strains of these points were calculated from the experimental measurement of the maximum depth at fracture by a theoretical procedure that was comprehensively described by the authors in their previous work [SIL 08]. As is seen, the agreement between the theory and experiments is good with regard to the fracture strains of the conical shapes that are formed under plane strain assumptions. Concerning the fracture strains at the corners of the pyramids, these are in good agreement with the FFLs but deviations occur between the experimental measurements and the estimates calculated from the experimental value of maximum depth under equal bi-axial stretching assumptions. The reason for this deviation may, once again, be attributed to the assumption of equal bi-axial stretching.

Figure 5.12 presents FE estimates of the meridional σ_ϕ and circumferential σ_θ stresses in three different locations placed along the inclined wall surface of a truncated conical shape. As is seen, the meridional stress σ_ϕ is lower than the yield stress and its value decreases along the inclined wall, being higher at the transition point C and smaller at point D (points C and D are defined in Figure 5.6). The circumferential stress σ_θ is close to zero and therefore can be neglected. This result confirms that the inclined wall surface of the sheet adjacent to the forming tool is elastic and further corroborates the applicability of the analytical framework developed by the authors.

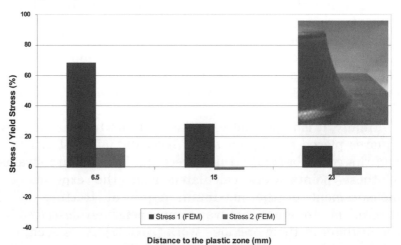

Figure 5.12. *FE estimates for the meridional (1) and circumferential (2) stresses in three different locations placed along the inclined wall surface of a truncated conical shape*

5.7. Examples of applications

The sheet metal parts shown in Figure 5.13 were selected to provide examples of the application of SPIF in various fields of engineering. The parts were produced in single or small batches and enabled significant savings in both raw material and energy when compared with alternative solutions based on conventional manufacturing technologies.

The first part (Figure 5.13a) is a prototype of a sector shower tray made from aluminum and its production process is comprehensively explained in the following section. The second, third, and fourth parts (Figures 5.13b, c, and d) are steel prototypes for the automotive industry, illustrating the advantages of SPIF in producing a sheet metal part that can be used directly in the function for which it is intended. The last two parts (Figures 5.13e and f) are biomedical, one-of-a-kind products in the form of a dental plate [TAN 05] and a cranial plate [DUF 06] of pure titanium.

5.7.1. *Sector shower tray*

Rapid prototyping of a sector shower tray produced by SPIF offers the challenge to manufacture a sheet metal part with vertical walls (Figure 5.13a). This is because according to the sine law, for a 90° drawing angle, the parts should have a final thickness equal to zero and strains approaching infinity.

The solution to overcome the aforementioned difficulties and obtain a sound sheet metal part requires the use of multi-stage SPIF. The first stage of the multi-stage SPIF process is used to produce the stripes and the circular depression shown in Figure 5.14a, whereas the subsequent stages, to be performed after turning over the sheet, are employed to obtain the top area (second stage) and gradually form the vertical walls (third to sixth stages).

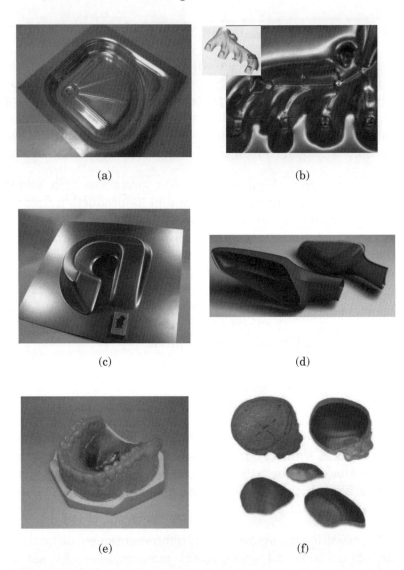

Figure 5.13. *Applications of SPIF: (a) sector shower tray; (b) automotive heat shield [JES 05]; (c) half part of an automotive exhaust silencer housing; (d) automotive side-view mirror consisting of three assembled components; (e) dental plate [TAN 05]; and (f) cranial plate [DUF 06]*

Single-Point Incremental Forming 205

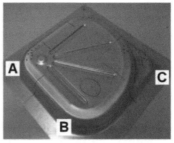

Figure 5.14. *SPIF of a sector shower tray: (a) prototype at the end of the first stage; (b) final part showing the circle grids that were used for measuring the strains at the bottom side of the part*

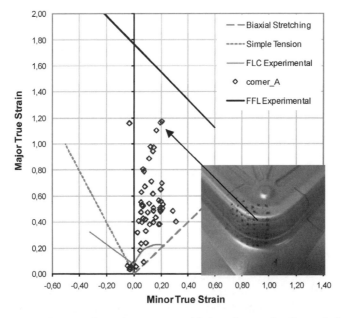

Figure 5.15. *Principal strain space containing the results from circle grid analysis for the corner of the sector shower tray labeled "A" in Figure 5.14*

The results from circle grid analysis for the most critical corner of the sector shower tray (labeled "A" in Figure 5.14) are shown in Figure 5.15. Strains are located well above the experimental FLC, but beneath the FFL, are in close agreement with the formability limits of the aluminum AA1050-O that were experimentally determined by means of tensile tests and circular and elliptical bulge tests.

5.8. Conclusions

This chapter has presented a theoretical framework of SPIF that is capable of dealing with the fundamentals of the process and explaining the experimental and numerical results that have been available in the literature for the past couple of years. The framework is based on membrane analysis with bi-directional, in-plane contact friction and is focused on the extreme modes of deformation that are likely to be found in SPIF processes [SIL 08]. The overall investigation is supported by numerical modeling and experimental techniques targeted at the construction and use of FLCs and FFLs.

Results confirm that stretching is the governing mode of deformation, that formability is limited by fracture without previous necking, and that the analytical framework is capable of successfully and effectively addressing the mechanics of deformation and the formability limits of the process.

The last part of the chapter dealt with small-scale production of sheet metal parts by single-stage and multi-stage SPIF. Rapid prototyping of a sector shower tray was presented as an example in which SPIF allows significant savings in both material and energy requirements when compared with conventional stamping.

5.9. References

[ALL 07] ALLWOOD J.M., SHOULER D.R., TEKKAYA A.E., "The increased forming limits of incremental sheet forming processes", *Key Engineering Materials*, 344, p. 621-628, 2007.

[ATK 96] ATKINS A.G., "Fracture in forming", *Journal of Materials Processing Technology*, vol. 56, p. 609-618, 1996.

[ATK 97] ATKINS A.G., "Fracture mechanics and metal forming: damage mechanics and the local approach of yesterday and today", in Rossmanith H.P. (Editor), *Fracture Research in Retrospect*, AA Balkema, Rotterdam, p. 327-350, 1997.

[AYA 84] AYADA M., HIGASHINO T., MORI K., "Central bursting in extrusion of inhomogeneous materials", in *Proceedings of ICTP 1984 International Conference on Technology of Plasticity*, Tokyo, Japan, 1984, Kudo H. (Editor), p. 553-558.

[CAO 08] CAO J., HUANG Y., REDDY N.V., MALHOTRA R., WANG Y., "Incremental sheet metal forming: advances and challenges", in *Proceedings of ICTP 2008 International Conference on Technology of Plasticity*, Gyeongju, Korea, 2008, Yang D.Y., Kim Y.H., Park C.H. (Editors), p. 751-752.

[DUF 06] DUFLOU J., *Production Process – Cranial Plate*, Katholieke Universiteit Leuven, 2006, http://www.mech.kuleuven.be/pe/pp/research/spif_cranial.en.html, November 2009.

[EMB 81] EMBURY J.D., DUNCAN J.L., "Formability maps", *Annual Review of Materials Science*, vol. 11, p. 505-521, 1981.

[EMM 08] EMMENS W. C., VAN DEN BOOGAARD A.H., "Incremental forming studied by tensile tests with bending", in *Proceedings of ICTP 2008 International Conference on Technology of Plasticity*, Gyeongju, Korea, 2008, Yang D.Y., Kim Y.H., Park C.H. (Editors), p. 245-246.

[EYC 07] EYCKENS P., HE S., VAN BAEL A., VAN HOUTTE P., DUFLOU J., "Forming limit predictions for the serrated strain paths in single point incremental sheet forming", in *Proceedings of NUMIFORM 2007*, Porto, Portugal, 2007, Cesar de Sa J.M.A., Santos A.D. (Editors), American Institute of Physics, p. 141-146.

[FIL 02] FILICE L., FRATINI L., MICARI F., "Analysis of material formability in incremental forming", *Annals of CIRP*, vol. 51, p. 199-202, 2002.

[HIR 06] HIRT G., AMES J., BAMBACH M., "Basic investigation into the characteristics of dies and support tools used in CNC-incremental sheet forming", in *Proceedings of the International Deep Drawing Research Group Conference IDDRG'06*, Porto, Portugal, 2006, p. 341-348.

[JAD 03] JADHAV S., GOEBEL R., HOMBERG W., KLEINER M., "Process optimization and control for incremental forming sheet metal forming", in *Proceedings of the International Deep Drawing Research Group Conference IDDRG'03*, Bled, Slovenia, 2003, p. 165-171.

[JES 01] JESWIET J., "Incremental single point forming", *Transactions of North American Manufacturing Research Institute*, vol. 29, p. 75-79, 2001.

[JES 05] JESWIET J., MICARI F., HIRT G., BRAMLEY A., DUFLOU J., ALLWOOD J., "Asymmetric single point incremental forming of sheet metal", *Annals of CIRP*, vol. 54, p. 623-650, 2005.

[KEG 61] KEGG R.L., "A new test method for determination of spinnability of metals", *Journal of Engineering for Industry, Transactions of ASME*, vol. 83, p. 119-124, 1961.

[KIT 96] KITAZAWA K., WAKABAYASHI A., MURATA K., YAEJIMA K., "Metal-flow phenomena in computerized numerically controlled incremental stretch-expanding of aluminum sheets", *Keikinzoku/Journal of Japan Institute of Light Metals*, vol. 46, p. 65-70, 1996.

[LES 67] LESZAK E., "Apparatus and process for incremental dieless forming", Patent US3342051A, 1967.

[MAT 01] MATSUBARA S., "A computer numerically controlled dieless incremental forming of a sheet metal", *Proceedings of the Institution of Mechanical Engineers, Part B, Journal of Engineering Manufacture*, vol. 215, p. 959-966, 2001.

[ROS 76] ROSSARD C., *Mise en forme des métaux et alliages*, CNRS, Paris, 1976.

[SIL 08] SILVA M.B., SKJOEDT M., ATKINS A. G., BAY N., MARTINS P.A.F., "Single point incremental forming & formability/failure diagrams", *Journal of Strain Analysis for Engineering Design*, vol. 43, p. 15-36, 2008.

[SKJ 09] SKJOEDT M., SILVA M.B., MARTINS P.A., BAY N., "Strategies and limits in multi-stage single-point incremental forming", *Journal of Strain Analysis for Engineering Design*, vol. 45, p. 1-12, 2009.

[TAN 05] TANAKA S., NAKAMURA T., HAYAKAWA K., NAKAMURA H., MOTOMURA K., "Incremental sheet metal forming process for pure titanium denture plate", in *Proceedings of the 8th International Conference on Technology of Plasticity ICTP'2005*, Verona, Italy, 2005, p. 135-136.

Chapter 6

Molding of Spent Rubber from Tire Recycling

Tire recycling is a growing problem that must be faced by the scientific community. However, economically feasible solutions must be found to ensure a sustainable development of the rubber market. Many interesting ideas are provided by the scientific literature even if many concerns are often present. The best idea deals with direct molding of rubber powders.

The vulcanization process is not reversible because of the rubber molecule cross-linking. By increasing the cross-linking density, stiffer materials are obtained with higher wear resistance. Molding rubber pellets or powders does not allow us to obtain the cross-linking density of the initial tire. As a consequence, the final mechanical and wear performances of the molded parts are expected to be lower. Even if the proposed recycling technology is not able to remanufacture new tires, many other industrial applications are possible if sufficient performances are achieved.

Chapter written by Fabrizio QUADRINI, Alessandro GUGLIELMOTTI, Carmine LUCIGNANO and Vincenzo TAGLIAFERRI.

6.1. Introduction

In modern industrial development, the enormous waste of materials and energy is becoming a serious social and economical problem. People have been made aware of the environmental risk connected with the industrial production, but there are not enough initiatives to reduce the related energy consumption or to extract raw materials from waste. Nowadays, the social and political attention on these themes is strong but the government agencies do not operate as efficiently as the scientific research. A typical example is given by the management of spent tires. The disposal of spent tires is already an increasing problem for the European Union where estimates are that around 250 million car and truck tires are scrapped each year, representing about 2.6 million tons of tires. Moreover, spent tires represent the main part of the total amount of the spent rubber material, which is produced every year in the world. In fact, of the annual total global production of rubber material, which amounts to 16–17 million tons, approximately 65% is used for the production of tires.

In the most industrialized countries, tires are usually dumped in landfills or left in the open air, with the consequence of significant environmental disturbances, such as non-biodegradable residues, and dangerous situations, such as the risk of fire. To face this problem, new regulations have been imposed which force the tire manufacturers to manage tire disposal, but these regulations do not solve the problem. Currently, the tire manufacturers are looking for the easiest solution with no regard for human safety or environmental conservation.

At present, the main solution is the energy reclaim by combustion of the rubber as there are some factories (e.g. cement kilns) that need high energies for their processes. The use of shredded tires for energy reclaim in cement kilns allows the tire manufacturers to solve the regulatory problem as the cement kilns are able to burn almost all the rubber coming from the spent tires, but it is questionable that this solution is sustainable or environmentally conscious. It is known that the rubber from spent tires cannot be used to produce other tires but it is not true that it cannot be used in any other manufacturing process. Even if it is politically correct, it seems to be ethically wrong to burn the rubber only to remove the tires from the landfills. Instead, scientific research is trying to propose new recycling technologies which would be able to add further manufacturing steps between the initial tire manufacturing and the final energy reclaiming. It is always possible to reclaim energy from the rubber by combustion and it would be environmentally correct to use this solution when other technological solutions are not practicable.

Figure 6.1. *Management of waste tires*

The case of spent tires is well known but not unique; many other materials pose serious problems for their recycling (such as glass fiber reinforced plastics of boats, carbon fiber reinforced plastics used for light-weight structures, or glass-filled phenolic resins used for electric insulators). All these materials as well as other materials (the rigid or flexible polyurethanes used in construction, the cross-linked polyethylene of thermo-shrinkable cables) share the same technological characteristic, being thermosets or behaving like them. Thermosets polymerize during the production process: as a result, a molecular network is generated which prevents the material flow under heating. For this reason, thermosets cannot be recycled by using the same technologies of thermoplastics (i.e. re-melting and re-plastification) and are generally considered rubbish at the end of their use. It is evident that thermoplastics should be preferred for everyday use but thermosets are mandatory when their unique properties cannot be guaranteed by any thermoplastic material. Thermosets are used in the field of transport or electric insulation where human safety is more important than the environmental impact. Therefore, the problem of their recycling will always exist and should be faced by the scientific community. Until now, solutions for thermoset recycling seem to be ineffective due to their high costs or emissions: only energy reclaim is considered a valid strategy even if it is a very old idea.

Factories are interested in developing industrial processes which provide high economical profits. The material recycling is felt by the factories as a cost and not as a business. If they exist, the recycling systems of thermosets are too expensive, large, or complex. In many cases, these systems are ineffective and very expensive research activities would be necessary to design an effective recycling technology. As a result, all the factories that produce thermosetting parts agree to their combustion for energy reclaim. Fortunately, environmental consciousness of the people is increasing, and the European policy is pressing to define a sustainable development of the industrial activities: the present limit is the absence of new technological solutions. In this chapter, the authors state that thermosets are high-performance materials which cannot become rubbish at the end of their life. Many other properties of these materials may be used apart from their ability to be burnt: structural and functional properties are useful in a lot of applications. A technological innovation is presented that could be able to fill the lack of knowledge about thermoset recycling. The idea at the basis of this new recycling technology is very simple: pulverizing materials to give new reactivity to the resulting powder. In fact, the broken links on the external surface of the particles may act as polymerizing sites in further processing steps. If a residual reactivity of the bulk material is also present, that is useful to increase the powder reactivity. Rubbers may be pulverized to produce a powder, which may be molded by compression molding without any addition of linking agents or virgin materials. As a result, a finished product would be obtained with good mechanical and functional properties. This product would also be recyclable by means of the same technology and energy reclaim is configured as the last step when further recycling steps are not technically feasible.

6.2. State of the art of tire recycling

In the last decade, a great effort has been made in the scientific community to solve the problem of tire recycling. Figure 6.1 summarizes some possible solutions in waste tire management. To have an outlook of the technical strategies for tire recycling, the scientific contribution in the last 5 years can be taken into consideration. Since 2004, numerous articles were published dealing with technologies for tire recycling.

Tire remanufacturing is the best way to manage spent tires. There are two competing technologies in tire retreading: the mold cure process and the pre-cure process [LEB 06]. In both cases, retreaded tires deliver the same mileage compared with new tires, although they are sold with discounts between 30% and 50%. However, there are several stages in the retreading process in which some material is lost: therefore, tire retreading is not always technically feasible. Moreover, this technology has reached its limits with respect to the fraction of the demand willing to buy "green tires". As a result, remanufacturing cannot be considered the only solution for spent tire recycling but only a valid alternative for small quantities of tires.

Tire grinding is generally the first step of any recycling process. Each tire is cut in to small parts and in such cases scrap tires are subsequently re-grinded to separate the spent rubber from the other tire components (steel and organic fibers). Crumb rubber can be produced from scrap tires with different particle sizes. The profitability of a crumb facility appears to be sensitive to crumb rubber prices, operating costs, and raw material availability [SUN 04]. The optimization of the grinding systems is already a matter of research: kinematics and wear of tool blades are essential to design an efficient process [SHI 04]. After grinding, cyclones and mechanical separators are used to remove tangled steel and rayon fibers: a vertical Venturi separator can be used as well [MCB 05].

Pyrolysis is a recycling process which uses crumb rubber. In tire pyrolysis, solid, liquid, and gaseous products are obtained. The solid compounds (about 40% weight of the processed rubber) are mostly made up of carbon black. In the latest studies, the characterization of liquid products from pyrolysis was deepened [LAR 04] as well as the pyrolysis kinetics under fast heating conditions [AGU 05]. Nitrogen thermal plasma pyrolysis has the advantage of eliminating the emission of toxic substances and the generation of liquid products [TAN 05, TAN 06]. However, many other technological solutions can be used to improve the process efficiency, such as a vertical reactor with multiple hollow discs [LIU 06], a conical spouted bed reactor [ARA 07], and a steam activation of the pyrolytic carbon black [MAS 07]. During pyrolysis, a non-condensable gas is formed which is a mixture of light hydrocarbons, carbon dioxide and monoxide, and hydrogen. Because of its high calorific value, this gas can be used to supply energy to the tire endothermic pyrolytic process, monitoring the related emissions [AYL 07]. Pyrolytic oils can also be burned for energy recovery [MUR 06].

Combustion is another way to recycle tires: shredded tires can be used directly without any successive grinding process. The production of energy and valuable chemical products from waste tires is the main goal of their chemical re-processing. In fact, catalytic pyrolysis and distillation lead to the production of an oil with high gross calorific value [SHA 07a]. Scrap tire-derived oils can also be upgraded by means of activated carbon-supported metal catalysts to produce liquid fuels [UCA 07]. Acetylene was obtained from the co-pyrolysis of biomass and waste tires [BAO 08]. Tire derived fuels (TDF) can be used a supplement fuel for the clinker production [PIP 05]. However, TDF cannot exceed 30% of the kiln fuel without adversely altering the chemistry of the cement's hardening process. As a further advantage, the use of TDF allows kilns to reduce NOx emissions [DAR 08]. TDF can be used also in cupolas [FTJ 06] or in fluidized beds for steam generation [DUO 07] and low-temperature gasification [XIA 08].

Activated carbons are widely used as adsorbents in gas-phase and liquid-phase separation processes. They are generally prepared from many carbonaceous materials including wastes and agricultural by-products. Gasification of tires with steam and carbon dioxide was also suggested to produce activated carbons from tires [GON 06], and fluidized beds can be used for this aim [MIN 06].

Dealing with their *absorption properties*, tire shreds can be directly used as drainage media in landfill leachate collection systems [WAR 05]. As tire rubber is flexible and hydrophobic (i.e. oil-philic), it is a good candidate as an oil adsorbent [LIN 08]. About 2.2 g of motor oil can be adsorbed to each gram of 20 mesh tire powder. Graft co-polymerization of tire rubber has also been used to improve the oil absorbency [WUB 09].

Chemical processes intend to extract energy and chemical substances from scrap tires. This way, the rubber microstructure has to be destroyed, with no respect for its residual performances. When tire re-manufacturing is not feasible, a structural application should always be preferred for the spent tire. In fact, structural parts made of recycled rubber directly reduce the consumption of natural resources and do not prevent successive chemical recycling.

The rubber of the spent tires still preserve good *damping properties*. Additives from recycled tires were used with plaster to change the acoustic and physical-mechanical properties of sound absorption materials [STA 07]. Recycled rubber granulates were used to damp rectangular tubes [CHE 08], and scrap tires were efficiently used as damping materials in the construction of roads. In the case of road embankments, using tire shreds as a lightweight fill can consume large quantities of scrap tires with certain engineering benefits [SHA 05]. Recycled tire chips allow to reduce railroad vibration due to transports [CHO 07] and to minimize dynamic earth pressure during compaction of backfill [LEE 07].

Asphalt mixtures represent another way to use spent tires for the construction of roads. In fact, the addition of recycled tire rubber in asphalt mixtures can improve their engineering properties in laboratory tests [CAO 07]. However, scrap tires in asphalt or other pavement applications, although technically viable, need to be subsidized in order to compete with conventional aggregates in meeting the technical requirements for asphalt pavements [HUA 07]. It was observed that modification of bitumen with rubber makes the mixture less compactable when compared with the mixtures made without rubber [CEL 08]. Moreover, the processing procedure and tire type are very important in the determination of the mixture viscosity [THO 09].

Concrete is another construction material under consideration for tire recycling. Many scientific studies deal with the production of rubber-modified concretes, but in this case the final results are not very impressive. From laboratory tests, it was observed that strength and stiffness of concrete modified with waste tire fibers or chips were always lower than those modified without waste tires [LIG 04]. Previous treatment of rubber with NaOH and silane does not produce significant changes on compressive strength and splitting tensile strength of Portland I concrete composites [ALB 05]. Self-compacting technology leads to better results [BIG 06], but the application of rubber granulates to replace coarse and fine aggregates in pedestrian concrete block seems to be more interesting [SUK 06]. In fact, crumb rubber can make concrete blocks more flexible and provide softness to the surface. Moreover, the material durability is acceptable [HER 07]. Alternatively, the typical properties of rubber-filled concrete can be exploited for producing controlled modulus columns [TUR 08] or improving thermal and sound properties of pre-cast panels [SUK 09].

Apart from concrete, re-used tire powders can be used as *fillers* for many other structural materials. The simplest case is the incorporation of rubber waste in virgin rubber. Partially devulcanized tire rubber can be revulcanized with different virgin rubbers [GRI 04]. Recycled tire materials were also used in natural rubber-based tread composites [BAN 06a] and in NR/BR blend-based tread composites [BAN 06b]. However, a deterioration in tensile strength, fatigue to failure, and abrasion properties were observed. In particular, it was found that using devulcanized rubber as part of rubber yields much better properties than using it as filler [LAM 06].

Structural composites can be produced by using non-rubber matrices. Granulated rubber and polyurethane pre-polymers were used for the synthesis of rubber waste-polyurethane composites [SUL 04]. Rubber powders were also modified with peroxide for producing polypropylene/rubber composites [SHA 07b]. However, the most interesting results were obtained by mixing rubber powders and polyethylene. It is particularly interesting to note the possibility of producing secondary materials by blending recycled polyethylene coming from greenhouses and tire rubber [SCA 05]. Avoiding the use of virgin materials seems to be the only successful recycling strategy, and polyethylene is present in a lot of civil and industrial applications. High-performance thermoplastic elastomers, based on recycled high-density polyethylene, olefinic-type ethylene-propylene-diene monomer rubber, and ground tire rubber treated with bitumen, were also prepared by using dynamic vulcanization technology [GRI 05], with a good compatibility between the components [GRI 06]. It was also observed that treatments with H_2SO_4 and silane coupling agent [COL 06], as well as γ irradiation [SON 06], improve the ability of rubber to interact with high-density polyethylene. Other methods to increase mechanical properties of composites are the addition of polypropylene filled with 30 wt% glass fiber [SHO 07]; the rubber oxidation [SON 07]; and the free radical mechanism by adding a peroxide [SON 08].

Rubber parts should be produced without any addition of virgin materials. Using additional treatments or virgin materials leads to an increase of the production cost, which is a serious obstacle to the development of a sustainable recycling technology. An interesting result was obtained by sintering rice husk-waste tire rubber mixtures: no linking agent was necessary even if the adhesion between the components has to be improved [GAR 07]. However, following the concept of a recycling technology without using virgin materials, the most interesting contribution was already given in 2003 by Bilgili *et al.* [BIL 03]. They proposed a new two-stage recycling process: first, the pulverization of the rubber granulates into small particles by means of a single screw extruder in the Solid State Shear Extrusion process; then, the compression molding of the produced powder in the absence of virgin rubber. Good mechanical performances were obtained at the end of the two-stage process. Bilgili *et al.* discussed that, after the pulverization, the single rubber particle acquired new reactivity because of the broken links on the external particle surface. It is very singular that among the latest contribution (last 5 years), almost no one deals with this recycling technology. A contribution can be cited regarding the preparation of composites from ground tire rubber and waste fiber by mechanical milling [ZHA 07]. A pan-mill mechanochemical reactor was developed to partly devulcanize the rubber which was subsequently mixed with the other components and revulcanized.

In a recent work, rubber particles from the mechanical grinding of tires were directly compression molded [GUG 09]. Thus, the Solid State Shear Extrusion process is not the only way to provide new reactivity to the recycled rubber particles. The proposed recycling technology was called "direct powder molding" so as to refer to the absence of any virgin material or linking agent.

6.3. Direct molding of rubber particles

Rubber samples were obtained by direct molding of pulverized and flake rubber from exhausted tires. The term "direct molding" is used to mean a compression molding without the addition of any binder or virgin rubber. Exhausted tires were comminuted by mechanical grinding to produce rubber powders with different size distribution. Tire grinding was carried out by Sycorex Ricerche Italia S.p.A. (Caserta, Italy), which produces sound absorbing sheets and tiles by adding a resin binder to the rubber particles. Figure 6.2 shows some details of the recycling implant. Three size distributions were available (namely fine, medium-sized, and coarse particles), and Figure 6.3 shows the powder size distributions extracted by sieving. The reported size is the dimension of the sieve mesh, and the highest value (2.5 mm) refers to the powder residual in the last sieve. A very large difference was observed between the finest and the coarsest particles (Sycorex Ricerche Italia S.p.A. [Caserta, Italy]).

Figure 6.2. *Tire grinding system*

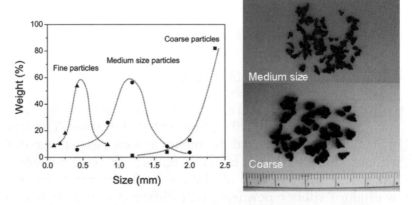

Figure 6.3. *Size distribution of particles from tire grinding*

To qualify the supplied material, thermal analysis was carried out by means of a Differential Scanning Calorimeter (Netzsch DSC 200PC). DSC tests were performed from –60°C to 250°C on a single rubber pellet: a double scan was carried out and the related curves are shown in Figure 6.4.

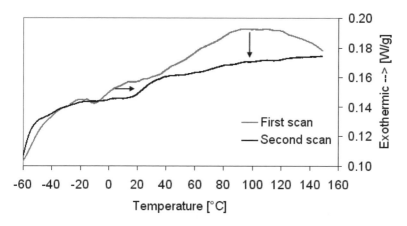

Figure 6.4. *DSC analyses on a rubber granule*

An important material modification occurred during the first heating: the material change is evident from the comparison between the first and the second DSC scan. A peak is present in the first scan near 100°C and disappears completely in the second scan. Moreover, an inflection point (probably related to a glass transition) is shifted from 0 to 20°C in the second heating. Those effects are probably dependent on the residual reactivity and the rubber re-structuration.

In fact, during the first heating, at high temperatures the rubber continues to polymerize, because of the residual reactivity, and an exothermic change of the DSC signal is visible. In the successive scan, the glass transition temperature of the rubber is higher, thanks to the higher cross-linking degree, and no other exothermic effect is visible.

Rubber parts were produced by compression molding of rubber powder without any addition of virgin rubber or linking agent (i.e. by direct molding). Thick quadrangular plates with a thickness about 20 mm, and the edge length of 150 mm, were molded in aluminum mold. The molding operation was carried out by means of a hot parallel plates press: an aluminum mold was used to produce large square pads. The choice of aluminum for the mold allowed the reduction of the process time in comparison with a steel mold having the same size, thanks to higher thermal conductivity and lower density of the aluminum alloy. The upper punch was particularly tall because of the high contraction of the powder under pressure during molding. The aluminum mold consisted of three parts with a frontal plate which could be removed to extract the molded pad. A hole in the mold allowed the insertion of a thermocouple to monitor the temperature during molding. The main disadvantage of this system was the limitation of the maximum pressure, which depended on the low strength of the threads that were machined in the mold to guarantee the closure.

In Figure 6.5 the aluminum mold, used for the molding tests, is shown together with some molded samples. All the samples were molded by means of a hydro-pneumatic press (by ATS FAAR) with a maximum load of 264 kN and the plate size of 300×300 mm^2. A low pressure was applied (2.6 MPa) because of the limited screw strength, and the plate temperature was fixed at 250°C. During the process, the material was left under the combined action of temperature and pressure until the value of 200°C was reached for the mold temperature. Afterwards, the plate heating was turned off and the mold was left to cool under pressure for following 20 min. This particular experimental procedure was necessary to reduce the molding time without generating material degradation.

Figure 6.5. *Aluminum mold and molded samples*

6.4. Experimental results

By increasing the molding time and pressure the rubber density increases. Even if the pressure effect is stronger, the best results are obtained with the combination of these two parameters. In fact, a good adhesion among the particles is obtained if a high contact pressure is applied for a long time. The temperature is another important factor because it defines the intrinsic mobility of the molecular segments of the rubber molecules. In order to reduce the process time, the temperature was set to the maximum value, which would avoid the material degradation. Moreover, if a residual reactivity is present in the comminuted material, polymerization occurs both in the powder bulk and in the contact surface among particles.

Because of the combined action of heat and pressure, the comminuted rubber was shaped in thick plates. Figure 6.6 shows some samples molded with the medium-size powder at different values of molding pressure. For each molding, the mold was filled with the same quantity of rubber powder (400 g), then it was placed between the hot parallel plates that were initially pre-heated at a temperature of 250°C: four pads were produced by applying four different pressure values (1.3, 1.95, 2.6 and 3.25 MPa) to investigate the effect of the pressure on mechanical and physical properties of the molded parts.

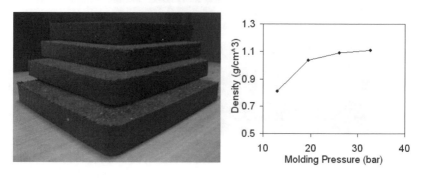

Figure 6.6. *Effect of the molding pressure on the molded rubber density*

A good moldability was observed for each molded condition, but the aesthetics part varied as a function of the molding pressure. Evidently, the higher contact pressure between the mold and the rubber particles, and among the same rubber particles, leads to the reduction of the porosity, both on the part surface and in the bulk. As expected, the plate density raised by increasing the molding pressure (Figure 6.6). The lower density value (0.8 g/cm^3) was related to the lower pressure value (1.30 MPa): by increasing the molding pressure up to 1.95 MPa, the density rapidly increased to 1.05 g/cm^3. However, further increments of the pressure led to small rises in density: the maximum density value was 1.1 g/cm³ at 3.25 MPa. This occurrence was confirmed by taking a look at the smoothness of the plate surface, which was very similar for the last three values of pressure and markedly worse for the lowest pressure.

Density is an important factor to evaluate the effectiveness of the molding process but it is not sufficient alone: valuable information about the adhesion among the rubber particles can be obtained by analyzing the tensile properties of the molded parts. Tensile tests are particularly suitable for evaluating the effectiveness of the powder molding process as they quantify the adhesion among the rubber particles. Five specimens were cut by sawing from each molded plate with a thickness ranging between 5 and 8 mm, a length equal to the plate edge length (150 mm), and a width equal to the molded plate thickness (20 mm). A gage length of 60 mm and a test rate of 10 mm/min were used: the test terminated with the sample rupture. Tests were performed by means of a universal material-testing machine (Alliance RT/50 by MTS) equipped with a load cell of 1 kN. Figure 6.7 shows some specimens cut from a molded plate together with the clamping system for tensile tests.

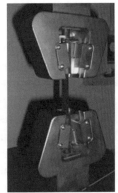

Figure 6.7. *Extraction of samples for tensile tests*

Figure 6.8 shows a typical tensile curve of a molded sample (100% fine particles) in comparison with the curve of a sample extracted from an industrial product made of rubber granulates with the addition of a linking agent (3 wt%, by Sycorex Ricerche Italia S.p.A.). The structure of the latter sample consists of soft rubber granules that are linked together by means of a rigid interface. The mechanical performance of the direct molded sample is very good, if considering the absence of any linking agent, as the tensile curves are comparable. Nevertheless, the part density is quite a bit higher.

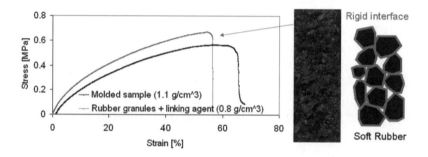

Figure 6.8. *Typical tensile curves of rubber products*

Tensile properties of molded samples of Figure 6.6 are reported in Figures 6.9a and 6.9b in terms of maximum stress and elongation at break, respectively. Tensile strength enhances by increasing the molding pressure (e.g. the part density). These results confirmed that mechanical properties are strongly dependent on the particle agglomeration. Generally, in tensile tests, a strength increase is related to a maximum elongation decrease. Instead, by increasing the molding pressure, the elongation at break increases as well as the tensile strength. This occurrence depends on the molded rubber structure where many bulk particles are linked together with small adhesion forces.

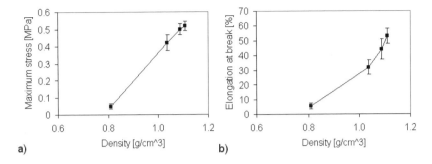

Figure 6.9. *Maximum stress (a) and elongation at break (b) of molded samples with different density*

The similarity of the tensile strength map and the elongation at break map is very singular as generally materials with higher strengths show lower elongations at break. For rubber pads, instead, increasing mechanical properties means increasing both strength and ductility. That is due to the intrinsic nature of the molded products which are made of granules joined together by their external surface. Sample break always occurs at the interface between particles and never inside the particles. Therefore, by improving the adhesion among particles, the ductility increases as well as the strength.

From tensile tests, the elastic modulus was also extracted. The elastic modulus seems to be constant in the molding pressure range between 1.95 and 3.25 MPa. In fact, it was expected that the elastic modulus is less dependent on the adhesion force among the particles in comparison with the tensile strength.

Further molding tests were carried out by using the different powder size distributions. Several mixtures of the available powders were prepared: rubber plates were molded with 100% of fine, medium-sized, or coarse particles, as well as with binary mixtures. All the powder mixtures showed a good moldability even if the part aesthetics and properties strongly depend on the powder size distribution as shown in Figure 6.10. Even if a good agglomeration was observed for all the powder mixtures, the shape of the coarse particles was always visible at the naked eye. The boundaries among the particles are easily recognizable, as the perfect joining cannot be achieved. Fine particles have to be used to obtain smooth surfaces.

Figure 6.10. *Molded samples with different powder size distribution*

Also other properties are greatly influenced by the powder distribution as well as the surface aspect. Generally, higher strength and elongation at break were achieved with finer size distributions. Moreover, elastic properties seem to be improved by the reduction of the average size. However, an optimum in the mechanical properties can be obtained by mixing the supplied powders.

Figure 6.11 shows the values of density and tensile properties for all the rubber powder mixtures. Binary mixtures were prepared by adding different content of fine powder to medium-sized or coarse particles. As Figure 6.11a shows, the molded rubber density is not dependent on the powder size distribution. Therefore, if a significant difference would be observed during mechanical testing, it is reasonable to assume that the difference depended on the particle linking. The density change is very low as the maximum difference is in the order of 10%. It is particularly important that similar densities can be obtained by using very different powder mixtures. However, the molded rubber density is not correlated to the mechanical performances as shown in Figure 6.11b in terms of tensile strength. The highest strength was measured in the case of 100% fine particles.

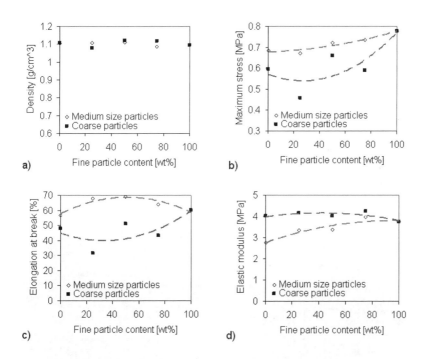

Figure 6.11. *Density (a), maximum stress (b), elongation at break (c), and elastic modulus (d) for binary mixtures of rubber powders*

In binary mixtures, higher strengths were obtained by using medium-sized particles. By decreasing the fine particle content, the tensile strength decreased both for the medium-sized and for the coarse particles. However, in the latter case a very high data scattering was observed.

Evidently, the low surface/volume ratio of the coarse particles affects the ultimate properties of the molded pads. The plot of the elongation at break (Figure 6.11c) is not very similar to the plot of the tensile strength (Figure 6.11b). Actually, a similar trend was visible for the coarse particles, but a maximum was observed for the medium-sized particles in the case of 50 wt% of fine particles.

It is important to observe that tensile testing is the hardest loading condition for the rubber samples (because of their granular structure) and much higher ultimate properties are expected in compression. The trend of the elastic modulus (Figure 6.11d), extracted from the same tensile tests, is very different from the discussed trends of maximum stress and elongation at break. In particular, higher values were obtained with the combination of coarse and fine particles. Probably, the presence of large granulates enhances the material stiffness even if the ultimate properties are poor.

Figure 6.12. *Complex and thick molded rubber pads*

In conclusion, Figure 6.12 shows a large rubber pad with a complex structure together with a very thick pad (about 40 mm of thickness). The pads were molded by using fine rubber powder and the mold of Figure 6.5.

6.5. Concluding remarks

Recycling of exhausted tires by compression molding of comminuted rubber is easy and cheap: the most accepted theory about this process describes the effect of the pressure and temperature in the following way. Broken links are present on the particle surfaces because of the comminution process. The pressure is necessary to put the particle interfaces into intimate contact, whereas the heating provides the energy to increase the molecular mobility; therefore, the broken links on the surface of the rubber particles originate new links, thereby "sintering" the particles into a single piece. Moreover, if a residual reactivity is present in the comminuted material, polymerization occurs in bulk powder and in the contact surface among particles. Another important effect of the temperature concerns the material softening which induces a higher contact among the particles under pressure. All the mentioned temperature effects are beneficial for the agglomeration of the rubber particles, but the temperature increase has a limit due to the occurrence of material burning or degradation.

Also for other thermosets, it would be possible to pulverize the materials and to directly mold the resulting powders without using any linking agent or virgin material. This process is very innovative in comparison with the technological state of the art in the field of recycling and is a necessary step to allow the sustainable development of the related manufacturing processes. In fact, the production of high-performance materials is developing very fast, and many new products are available on the market everyday. As a result, recycling procedures are generally not suitable for this kind of material. Generally, thermosets are considered exhausted materials when the part, which is made of it, is no longer working. Instead, exhausted thermosets should be considered a resource which can be used for other production processes.

6.6. References

[AGU 05] AGUADO R., OLAZAR M., VELEZ D., ARABIOURRUTIA M., BILBAO J., "Kinetics of scrap tyre pyrolysis under fast heating conditions", *Journal of Analytical Applied and Pyrolysis*, vol. 73, p. 290-298, 2005.

[ALB 05] ALBANO C., CAMACHO N., REYES J., FELIU J.L., HERNANDEZ M., "Influence of scrap rubber addition to Portland I concrete composites: destructive and non-destructive testing", *Composite Structures*, vol. 71, p. 439-446, 2005.

[ARA 07] ARABIOURRUTIA M., LOPEZ G., ELORDI G., OLAZAR M., AGUADO R., BILBAO J., "Product distribution obtained in the pyrolysis of tyres in a conical spouted bed reactor", *Chemical Engineering Science*, vol. 62, p. 5271-5275, 2007.

[AYL 07] AYLON E., MURILLO R., FERNANDEZ-COLINO A., ARANDA A., GARCIA T., CALLEN M.S., MASTRAL A.M., "Emission from the combustion of gas-phase products at tyre pyrolysis", *Journal of Analytical Applied Pyrolysis*, vol. 79, p. 210-214, 2007.

[BAN 06a] BANDYOPADHYAY S., AGRAWAL S.L., MANDOT S.K., MANDAL N., DASGUPTA S., MUKHOPADHYAY R., DEURI A.S., AMETA S.C., "Use of recycled tyre material in natural rubber-based tyre tread cap compound: part I", *Progress in Rubber Plastics Recycling Technology*, vol. 22, p. 45-60, 2006.

[BAN 06b] BANDYOPADHYAY S., DASGUPTA S., AGRAWAL S.L., MANDOT S.K., MANDAL N., MUKHOPADHYAY R., DEURI A.S., AMETA S.C., "Use of recycled tyre material in NR/BR blend based tyre tread compound: part II (with ground crumb rubber)", *Progress in Rubber Plastics Recycling Technology*, vol. 22, p. 269-284, 2006.

[BAO 08] BAO W., CAO Q., LV Y., CHANG L., "Acetylene from the co-pyrolysis of biomass and waste tires or coal in the H2/Ar plasma", *Energy Sources*, vol. 30, no. 5-8, p. 734-741, 2008.

[BIG 06] BIGNOZZI M.C., SANDROLINI F., "Tyre rubber waste recycling in self compacting-concrete", *Cement and Concrete Research*, vol. 36, p. 735-739, 2006.

[BIL 03] BILGILI E., DYBEK A., ARASTOOPOUR H., BERNSTEIN B., "A new recycling technology: compression molding of pulverized rubber waste in the absence of virgin rubber", *Journal of Elastomers and Plastics*, vol. 35, p. 235-256, 2003.

[CAO 07] CAO W., "Study on properties of recycled tire rubber modified asphalt mixtures using dry process", *Construction and Building Materials*, vol. 21, p. 1011-1015, 2007.

[CEL 08] CELIK O.N., ATIS C.D., "Compactibility of hot bituminous mixtures made with crumb rubber-modified binders", *Construction and Building Materials*, vol. 22, p. 1143-1147, 2008.

[CHE 08] CHETTAH A., CHEDLY S., ICHCHOU M., BAREILLE O., ONTENIENTE J.P., "Experimental and numerical investigation of flexural vibration damping by recycled rubber granulates", *International Journal of Vehicle Noise and Vibration*, vol. 4, p. 70-92, 2008.

[CHO 07] CHO S.D., KIM J.M., LEE K.W., "Utilization of waste tires to reduce railroad vibration", *Materials Science Forum*, vol. 544-545, p. 637-640, 2007.

[COL 06] COLOM X., CANAVATE J., CARRILLO F., VELASCO J.I., PAGES P., MUJAL R., NOGUES F., "Structural and mechanical studies on modified reused tyres composites", *European Polymer Journal*, vol. 42, p. 2369-2378, 2006.

[DAR 08] DARABI P., YUAN J., SALUDEAN M., "Numerical studies of mid-kiln tyre combustion", *Advances in Cement Research*, vol. 20, no. 3, p. 121-128, 2008.

[DUO 07] DUO W., KARIDIO I., CROSS L., ERICKSEN B., "Combustion and emission performance of a hog fuel fluidized bed boiler with addition of tire derived fuel", *Journal of Energy Resources Technology*, vol. 129, no. 1, p. 42-49, 2007.

[FTJ 06] "Beneficial reuse of tyres in cupolas", *Foundry Trade Journal*, vol. 180, no. 3635, p. 164-165, 2006.

[GAR 07] GARCIA D., LOPEZ J., BALART R., RUSECKAITE R.A., STEFANI P.M., "Composites based on sintering rice husk–waste tire rubber mixtures", *Materials and Design*, vol. 28, p. 2234-2238, 2007.

[GON 06] GONZALEZ J.F., ENCINAR J.M., GONZALEZ-GARCIA C.M., SABIO E., RAMIRO A., CANITO J.L., GANAN J., "Preparation of activated carbons from used tyres by gasification with steam and carbon dioxide", *Applied Surface Science*, vol. 252, p. 5999-6004, 2006.

[GRI 04] GRIGORYEVA O., FAINLEIB A., STAROSTENKO O., DANILENKO I., KOZAK N., DUDARENKO G., "Ground tire rubber (GTR) reclamation: virgin rubber/reclaimed GTR (RE)vulcanizates", *Rubber Chemistry and Technology*, vol. 77, no. 1, p. 131-146, 2004.

[GRI 05] GRIGORYEVA O.P., FAINLEIB A.M., TOLSTOV A.L., STAROSTENKO O.M., LIEVANA E., KARGER-KOCSIS J., "Thermoplastic elastomers based on recycled high-density polyethylene, ethylene–propylene–diene monomer rubber, and ground tire rubber", *Journal of Applied Polymer Science*, vol. 95, p. 659-671, 2005.

[GRI 06] GRIGORYEVA O., FAINLEIB A.M., TOLSTOV A., PISSIS P., SPANOUDAKI A., VATALIS A., DELIDES C., "Thermal analysis of thermoplastic elastomers based on recycled polyethylenes and ground tyre rubber", *Journal of Thermal Analysis and Calorimetry*, vol. 86, p. 229-233, 2006.

[GUG 09] GUGLIELMOTTI A., LUCIGNANO C., QUADRINI F., "Production of rubber pads by tire recycling", *International Journal of Materials Engineering Innovation*, vol. 1, no. 1, p. 91-106, 2009.

[HER 07] HERNANDEZ-OLIVARES F., BARLUENGA G., PARGA-LANDA B., BOLLATI M., WITOSZEK B., "Fatigue behaviour of recycled tyre rubber-filled concrete and its implications in the design of rigid pavements", *Construction and Building Materials*, vol. 21, p. 1918-1927, 2007.

[HUA 07] HUANG Y., BIRD R.N., HEIDRICH O., "A review of the use of recycled solid waste materials in asphalt pavements", *Resources, Conservation and Recycling*, vol. 52, p. 58-73, 2007.

[LAM 06] LAMMINMAKI J., LI S., HANHI K., "Feasible incorporation of devulcanized rubber waste in virgin natural rubber", *Journal of Materials Science*, vol. 41, no. 21, p. 8301-8307, 2006.

[LAR 04] LARESGOITI M.F., CABALLERO B.M., DE MARCO I., TORRES A., CABRERO M.A., CHOMON M.J., "Characterization of the liquid products obtained in tyre pyrolysis", *Journal of Analytical and Applied Pyrolysis*, vol. 71, p. 917-934, 2004.

[LEB 06] LEBRETOM B., TUMA A., "A quantitative approach to assessing the profitability of car and truck tire remanufacturing", *International Journal of Production Economics*, vol. 104, no. 2, p. 639-652, 2006.

[LEE 07] LEE H.J., ROH H.S., "The use of recycled tire chips to minimize dynamic earth pressure during compaction of backfill", *Construction and Building Materials*, vol. 21, p. 1016-1026, 2007.

[LIG 04] LI G., GARRICK G., EGGERS J., ABADIE C., STUBBLEFIELD M.A., PANG S., "Waste tire fiber modified concrete", *Composites: Part B*, vol. 35, p. 305-312, 2004.

[LIN 08] LIN C., HUANG C.L., SHERN C.C., "Recycling waste tire powder for the recovery of oil spills", *Resources, Conservation and Recycling*, vol. 52, p. 1162-1166, 2008.

[LIU 06] LIU B.Q., JIANG J.L., "Mathematical model for mass and heat transfer in multilevel reactor for pyrolysis of used tyres", *Journal of the Energy Institute*, vol. 79, p. 180-186, 2006.

[MAS 07] MASTRAL A.M., ARANDA A., MURILLO R., GARCIA T., CALLEN M.S., "Steam activation of tyre pyrolytic carbon black: kinetic study in a thermobalance", *Chemical Engineering Journal*, vol. 126, p. 79-85, 2007.

[MCB 05] MCBRIDE W., KEYS S., "Application of a vertical venturi separator for improved recycling of automotive tyres", *Chemical Engineering and Processing*, vol. 44, p. 287-291, 2005.

[MIN 06] MIN A., HARRIS A.T., "Influence of carbon dioxide partial pressure and fluidization velocity on activated carbons prepared from scrap car tyre in a fluidized bed", *Chemical Engineering Science*, vol. 61, p. 8050-8059, 2006.

[MUR 06] MURILLO R., AYLON E., NAVARRO M.V., CALLEN M.S., ARANDA A., MASTRAL A.M., "The application of thermal processes to valorise waste tyre", *Fuel Processing Technology*, vol. 87, p. 143-147, 2006.

[PIP 05] PIPILIKAKI P., KTSIOTI M., PAPAGEORGIOU D., FRAGOULIS D., CHANIOTAKIS E., "Use of tire derived fuel in clinker burning", *Cement & Concrete Composites*, vol. 27, p. 843-847, 2005.

[SCA 05] SCAFFARO R., DINTCHEVA T., NOCILLA M.A., LA MANTIA F.P., "Formulation, characterization and optimization of the processing condition of blends of recycled polyethylene and ground tyre rubber: mechanical and rheological analysis", *Polymer Degradation and Stability*, vol. 90, p. 281-287, 2005.

[SHA 05] SHALABY A., KHAN R., "Design of unsurfaced roads constructed with large-size shredded rubber tires: a case study", *Resource Conservation & Recycling*, vol. 44, p. 318-332, 2005.

[SHA 07a] SHAH J., JAN M.R., MABOOD F., "Catalytic conversion of waste tyres into valuable hydrocarbons", *Journal of Polymers and the Environment*, vol. 15, no. 3, p. 207-211, 2007.

[SHA 07b] SHANMUGHARAJ A.M., KIM J.K., RYU S.H., "Modification of rubber powder with peroxide and properties of polypropylene/rubber composites", *Journal of Applied Polymer Science*, vol. 104, p. 2237-2243, 2007.

[SHI 04] SHIH A.J., MCCALL R.C., "Kinematics and wear of tool blades for scrap tire shredding", *Machining Science and Technology*, vol. 8, no. 2, p. 193-210, 2004.

[SHO 07] SHOJAEI A., YOUSEFIAN H., SAHARKHIZ S., "Performance characterization of composite materials based on recycled high-density polyethylene and ground tire rubber reinforced with short glass fibers for structural applications", *Journal of Applied Polymer Science*, vol. 104, p. 1-8, 2007.

[SON 06] SONNIER R., LEROY E., CLERC L., BERGERET A., LOPEZ-CUESTA J.M., "Compatibilisation of polyethylene/ground tyre rubber blends by γ irradiation", *Polymer Degradation and Stability*, vol. 91, p. 2375-2379, 2006.

[SON 07] SONNIER R., LEROY E., CLERC L., BERGERET A., LOPEZ-CUESTA J.M., "Polyethylene/ground tyre rubber blends: influence of particle morphology and oxidation on mechanical properties", *Polymer Testing*, vol. 26, p. 274-281, 2007.

[SON 08] SONNIER R., LEROY E., CLERC L., BERGERET A., LOPEZ-CUESTA J.M., BRETELLE A.S, IENNY, P., "Compatibilizing thermoplastic/ground tyre rubber powder blends: efficiency and limits", *Polymer Testing*, vol. 27, p. 901-907, 2008.

[STA 07] STANKEVICIUS V., SKIRPKIUNAS G., GRINYS A., MISKINIS K., "Acoustical characteristics and physical-mechanical properties of plaster with rubber waste additives", *Material Science*, vol. 13, no. 4, p. 304-309, 2007.

[SUK 06] SUKONTASUKKUL P., CHAIKAEW C., "Properties of concrete pedestrian block mixed with crumb rubber", *Construction and Building Materials*, vol. 20, p. 450-457, 2006.

[SUK 09] SUKONTASUKKUL P., "Use of crumb rubber to improve thermal and sound properties of pre-cast concrete panel", *Construction and Building Materials*, vol. 23, p. 1084-1092, 2009.

[SUL 04] SULKOWSKI W.W., DANCH A., MOCZYNSKI M., RADON A., SULKOWSKA A., BOREK J., "Thermogravimetric study of rubber waste-polyurethane composites", *Journal of Thermal Analysis and Calorimetry*, vol. 78, no. 3, p. 905-921, 2004.

[SUN 04] SUNTHONPAGASIT N., DUFFEY M.R., "Scrap tires to crumb rubber: feasibility analysis for processing facilities", *Resources Conservation & Recycling*, vol. 40, p. 281-299, 2004.

[TAN 05] TANG L., HUANG H., "Thermal plasma pyrolysis of used tires for carbon black recovery", *Journal of Material Science*, vol. 40, p. 3817-3819, 2005.

[TAN 06] TANG L., HUANG H., "Pyrolysis processing of used old tyres by thermal plasma technology", *International Journal of Environmental Technology and Management*, vol. 6, p. 631-640, 2006.

[THO 09] THODESEN C., SHATANAWI K., AMIRKHANIAN S., "Effect of crumb rubber characteristics on crumb rubber modified (CRM) binder viscosity", *Construction and Building Materials*, vol. 23, p. 295-303, 2009.

[TUR 08] TURATSINZE A., GARROS M., "On the modulus of elasticity and strain capacity of self-compacting concrete incorporating rubber aggregates", *Resources, Conservation and Recycling*, vol. 52, p. 1209-1215, 2008.

[UCA 07] UCAR S., KARAGOZ S., YANIK J., YUKSEL M., SAGLAM M., "Upgrading scrap tire derived oils using activated carbon supported metal catalysts", *Energy Sources*, vol. 29, no. 4-8, p. 425-437, 2007.

[WAR 05] WARITH M.A., EVGIN E., BENSON P.A.S., RAO S.M., "Evaluation of permeability of tire shreds under vertical loading", *Journal of Testing and Evaluation*, vol. 33, no. 1, p. 51-54, 2005.

[WUB 09] WU B., ZHOU M.H., "Recycling of waste tyre rubber into oil absorbent", *Waste Management*, vol. 29, p. 355-359, 2009.

[XIA 08] XIAO G., NI M.J., CEN K.F., "Low-temperature gasification of waste tire in a fluidized bed", *Energy Conversion and Management*, vol. 49, p. 2078-2082, 2008.

[ZHA 07] ZHANG X.X., LU C.H., LIANG M., "Preparation of rubber composites from ground tire rubber reinforced with waste-tire fiber through mechanical milling", *Journal of Applied Polymer Science*, vol. 103, p. 4087-4094, 2007.

List of Authors

Paulo Roberto de AGUIAR
Department of Mechanical Engineering
UNESP
Brazil

Viktor P. ASTAKHOV
Department of Mechanical Engineering
Michigan State University
East Lansing
USA

Niels BAY
Technical University of Denmark
Department of Mechanical Engineering
Kgs. Lyngby
Denmark

Eduardo Carlos BIANCHI
Department of Mechanical Engineering
UNESP
Brazil

Rubens Chinali CANARIM
Department of Mechanical Engineering
UNESP
Brazil

J. Paulo DAVIM
Department of Mechanical Engineering
University of Aveiro
Portugal

Vinayak N. GAITONDE
B.V.B. College of Engineering and Technology
Department of Industrial and Production Engineering
Karnataka
India

Alessandro GUGLIELMOTTI
Department of Mechanical Engineering
University of Rome Tor Vergata
Italy

Ramesh S. KARNIK
B.V.B. College of Engineering and Technology
Department of Electrical and Electronic Engineering
Karnataka
India

Jong-Leng LIOW
School of Engineering and Information Technology
University of New South Wales
Canberra
Australia

Carmine LUCIGNANO
Department of Mechanical Engineering
University of Rome Tor Vergata
Italy

Paulo A. F. MARTINS
IDMEC
Instituto Superior Técnico
Technical University of Lisbon
Portugal

Fabrizio QUADRINI
Department of Mechanical Engineering
University of Rome Tor Vergata
Italy

Maria Beatriz SILVA
IDMEC
Instituto Superior Técnico
Technical University of Lisbon
Portugal

Leonardo Roberto da SILVA
Department of Materials Engineering
CEFET
Belo Horizonte
Brazil

Vincenzo TAGLIAFERRI
Department of Mechanical Engineering
University of Rome Tor Vergata
Italy

Index

A, B

ANN modeling, 79, 91, 92, 94, 104
Asphalt mixtures, 218, 235
Bernstein distribution, 33, 59, 60, 62, 66

C

Concrete, 219, 234–236, 238–240
Cutting fluids, 79, 80, 82, 85, 105, 108–110, 112–115, 120, 121
Cutting tool reliability, 33, 37–39, 44–46, 67
Cutting tool sustainability, 33

D

Design of MQL systems, 116
Direct molding, 211, 221, 224
Drilling, 82, 84, 85, 89, 104–107, 109, 110
Dry cutting, 79, 81, 86, 87, 108

E, F

Energy, 1–4, 9, 14, 22, 23, 25, 29
Environmental impact, 1, 5, 7
Finite element, 173
Formability limits, 173, 179, 181, 185, 196, 206, 199
Forming limit diagrams, 191
Fracture forming limit diagrams, 191

G, I, L

Grinding, 111–113, 117–121, 123–126, 128–130, 132–137, 140, 142, 146, 153–157, 159–165, 167–172
Incremental forming with counter tool (IFCT), 174
Incremental sheet forming processes, 174, 206
Internal plunge grinding, 146
Life cycle analysis (LCA), 1

M

Machinability of brass, 91, 99, 107
Machining, 79–81, 83–87, 90, 92, 104–110
Machining processes, 79, 84, 89, 90, 104
Manufacturing methods, 5, 7, 10, 12
Mass balance, 9, 14
Mass energy, 9, 14
Membrane analysis, 173, 181, 183, 190, 206
Micro-device manufacturing, 1, 4–7, 10, 11, 13, 16, 28, 31
Micro-end-milling, 1, 5, 12, 22, 24–26, 28
Micro-manufacturing, 5, 9, 10, 14
Milling, 82, 86, 87, 89, 104, 107, 108–110
Minimum quantity lubrication (MQL), 79, 105–108, 110, 111
Minimum work, 16–21, 27, 28
Molding of spent rubber, 211

P, R

Plunge external cylindrical grinding, 122
Probability density function, 41, 42, 50
Pyrolysis, 216, 217, 234, 235, 237, 239
Rubber particles, 221, 226, 227, 233

S

Single-point incremental forming (SPIF), 173
Statistical analysis, 52, 54, 61
Stress and strain, 182, 185, 196
Structural composites, 219
Surface grinding, 119, 154
Sustainability, 1, 5, 31, 33, 35, 37, 39, 50, 77

T

Taguchi approach, 99
Tire grinding, 216, 221, 222
Tire recycling, 211, 215, 216, 219, 236
Tire remanufacturing, 215, 237
Tool flank wear, 41, 42, 48–50, 72
Tool life, 36, 37, 42, 45, 48–51, 53, 58–60, 63, 64, 71, 72, 74, 75, 77
Tool quality, 38, 39, 54, 58, 76
Tool wear curves, 52, 62
Turning, 79, 81, 82, 87–89, 91, 92, 95, 99, 101, 104–110
Two-point incremental forming (TPIF), 174